CE

13+

Science

Exam Practice
Questions
and Answers

GALORE
PARK

AN HACHETTE UK COMPANY

About the author

Ron Pickering has published a number of very successful books covering the GCSE, IGCSE and A level specifications and has worked in both maintained and independent education for more than 30 years. He now divides his time between teacher training, both in the UK and overseas, and writing, and has been a science advisor and curriculum manager at Altrincham Grammar School for Girls, as well as a Science Inspector for OFSTED.

Ron extends his interest in science by spending many hours photographing animals, both in the wild and in captive environments, as well as pondering on the power of steam as a guide on the North York Moors Railway. He tries to maintain some level of fitness by cycling the endless hills of North Yorkshire.

Dedication

I dedicate this book to all young scientists, wherever they are, but especially to three microscientists: Noah and Kay, our beloved grandsons, and Nola, their extremely inquisitive cousin.

Acknowledgements

Every effort has been made to trace all copyright holders, but if any have been inadvertently overlooked, the Publishers will be pleased to make the necessary arrangements at the first opportunity.

Although every effort has been made to ensure that website addresses are correct at time of going to press, Galore Park cannot be held responsible for the content of any website mentioned in this book. It is sometimes possible to find a relocated web page by typing in the address of the home page for a website in the URL window of your browser.

Hachette UK's policy is to use papers that are natural, renewable and recyclable products and made from wood grown in well-managed forests and other controlled sources. The logging and manufacturing processes are expected to conform to the environmental regulations of the country of origin.

Orders: **Teachers** please contact Hachette UK Distribution, Hely Hutchinson Centre, Milton Road, Didcot, Oxfordshire, OX11 7HH. Telephone: (44) 01235 400555.
Email: primary@hachette.co.uk. Lines are open from 9 a.m. to 5 p.m., Monday to Friday

Parents, **Tutors** please call: 020 3122 6405 (Monday to Friday , 9:30 a.m. to 4:30 p.m.).
Email: parentenquiries@hachette.co.uk

Visit our website at www.galorepark.co,uk for details of other revision guides for Common Entrance, examination papers and Galore Park publications.

ISBN: 978 1 3983 26507

Illustrations by Aptara, Inc.

Typeset in India

Printed and bound by CPI Group (UK) Ltd, Croydon, CR0 4YY

A catalogue record for this title is available from the British Library.

FSC
www.fsc.org
MIX
Paper | Supporting
responsible forestry
FSC™ C104740

Contents

Introduction

Common Entrance 13+ Science Exam Practice Questions and Answers is a book of sample exercises and their answers, for Common Entrance preparation, based on the ISEB specification for Science.

Unlike a Common Entrance exam paper, each chapter of the book tests a single topic. The aim of this structure is to allow you to focus on the topics in which you feel you are weakest, reinforcing your understanding of key terms, as well as your knowledge of the relevant ideas.

Just as in the Common Entrance papers, each section begins with a series of multiple-choice questions. These questions will very quickly tell you whether you know the basics of the test topic. The questions include a variety of styles typical of Common Entrance examinations at both the Foundation level and Level 2.

Higher level content which is not on the 13+ specification is indicated by a grey background tint like this paragraph.

Timing

Try to complete each test within 40 minutes, which is the time allocated to each Level 2 exam paper (Biology, Chemistry and Physics). Do not count the time you spend copying graphs or tables.

If you are entitled to extra time, use it according to the advice from your teacher, who will also advise you if 40 minutes is not the appropriate time for any particular test.

Drawing graphs

You should write all of your answers on separate paper (not in the book), and this includes drawing graphs. Where a graph is provided in the book, copy it (including axes, numbers and labels) onto appropriate graph paper and draw points and lines as required.

Calculations

CE papers require you to show your working when you answer a numerical question. This can be very helpful, as you may be awarded marks for working even if your final answer is incorrect.

Straight lines

Use a ruler to draw straight lines, for example when drawing rays of light or putting a border on a table of results.

Your exams at 13+

Assessment of the 13+ syllabus can occur at two levels: Foundation and Level 2. The syllabus is common for both levels. Candidates who are expected to achieve less than an average of 40% on the three Level 2 papers should consider using the Foundation paper.

Foundation (80 marks; 60 minutes)

There will be one paper with approximately equal numbers of questions based on the 13+ Biology, Chemistry and Physics syllabuses. The paper will consist of a mixture of closed items, for example multiple choice, matching pairs, completing sentences and some open questions. Open questions will have several parts, some of which will require answers of one or two sentences. These parts will carry a maximum of 3 marks. At least 25% of the paper will be testing 'Thinking and working as a scientist'.

For questions that require the use of formulae, equations will be provided.

Rearrangement of equations will not be required.

There will be no choice of questions. The use of calculators and protractors will be allowed in the examination.

Level 2 (60 marks; 40 minutes per paper)

There will be three papers, one in each of Biology, Chemistry and Physics. Some of the questions may be closed, although most will be open, with several parts requiring candidates to answer in sentences. These parts will carry a maximum of 4 marks. The maximum number of marks per question will be 12. At least 25% of the paper will be testing Thinking and working as a scientist.

There will be no choice of questions. The use of calculators and protractors will be allowed in the examination.

For quantitative questions that require the use of formulae, equations given in the syllabus will not be provided.

Scholarship (90 minutes)

Scholarship papers are based on the ISEB 13+ specification. The Common Academic Scholarship Examination (90 minutes, including 10 minutes of reading time) will be divided into three sections: A (Biology), B (Chemistry) and C (Physics). Candidates will be required to attempt all questions. Each section is worth 25 marks but the number of questions will vary. The use of calculators and protractors will be allowed in the examination.

For quantitative questions that require the use of formulae, equations given in the syllabus will not be provided. Rearrangement of equations may be required.

Know what to expect in the examination

- Use past papers to familiarise yourself with the format of the exam.
- Make sure you understand the language examiners use.

Before the examination

- Have all your equipment and pens ready the night before.
- Make sure you are at your best by getting a good night's sleep before the exam.
- Have a good breakfast in the morning.
- Take some water into the exam if you are allowed.
- Think positively and keep calm.

During the examination

- Have a watch on your desk. Work out how much time you need to allocate to each question and try to stick to it.
- Make sure you read and understand the instructions and rules on the front of the exam paper.
- Allow some time at the start to read and consider the questions carefully before writing anything.
- Read all the questions at least twice. Don't rush into answering before you have a chance to think about it.
- If a question is particularly hard, move on to the next one. Go back to it if you have time at the end.
- Always look on the back of the paper because there might be questions there that you have not answered.
- Check your answers make sense if you have time at the end.

Tips for the science examination

- You should write your answers on the question paper; you may use a calculator and remember all questions should be attempted.
- Look at the number of marks allocated for each question in order to assess how many relevant points are required for a full answer. Very often, marks are awarded for giving your reasons for writing a particular answer.
- In numerical questions, working out should be shown and the correct units used.
- Practical skills are important. Look back in your lab notes to remind yourself about why you carried out any practical work. What were you trying to find out? What did you actually do? What instrument did you use to take any measurements and what units did you use? How did you record your results: tables, bar charts or graphs? What were the results of your investigation and did you make any plans to change or improve what you did? These are all important and will be tested in the examination.
- A thorough understanding of your practical work will also help you to remember the key facts by putting them into context.

Exam Practice Questions
Experiments in science

1   Choose the option which best completes each of the following.

   a)   A colourless gas that turns limewater milky is _____   (1)

   oxygen   **hydrogen**

   carbon monoxide   **carbon dioxide**

   b)   Anhydrous copper sulfate can be used to test for water. If water is present, anhydrous
        copper sulfate turns from _____   (1)

   blue to white   **blue to pink**

   blue to black   **white to blue**

   c)   A colourless gas that will relight a glowing splint is _____   (1)

   oxygen   **hydrogen**

   carbon monoxide   **carbon dioxide**

   d)   The hottest part of a Bunsen flame is _____   (1)

   bright yellow   **deep blue**

   pale blue   **red-orange**

   e)   A factor that a student chooses to change during the course of an experiment is

        _____   (1)

   a fixed variable   **a controlled variable**

   an independent variable   **a dependent variable**

   f)   A gas that is released when zinc reacts with hydrochloric acid burns with a 'pop'.
        The gas is _____   (1)

   chlorine   **hydrogen**

   oxygen   **carbon dioxide**

   g)   A piece of apparatus used to measure and transfer small volumes of
        liquids is _____   (1)

   a burette   **a measuring cylinder**

   a pipette   **an evaporating dish**

2   The diagram below shows six pieces of laboratory equipment.

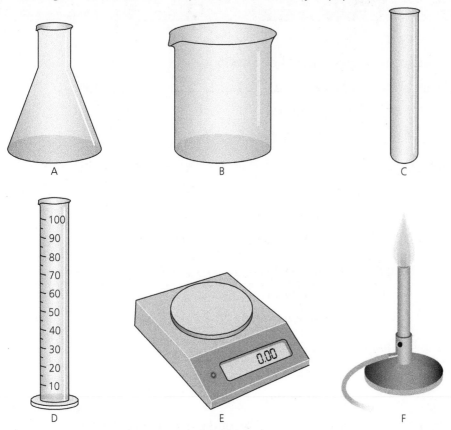

A   B   C

D   E   F

a) Lucy investigates how quickly sugar dissolves in water.

    i)   State which piece of equipment she would use to weigh 10 g of sugar.   (1)

    ii) State which piece of equipment she would use to measure out 90 cm³ of water.   (1)

b) Lucy heats the water in a beaker.

    i)   Suggest **one** safety precaution Lucy should take when heating the water.   (1)

    Lucy adds the 10 g of sugar to the hot water and measures the time taken for the sugar
    to dissolve. The equipment used to measure the time taken is not shown in the diagram.

    ii) State the name of the piece of equipment used to measure the time taken.   (1)

    iii) State the name of a piece of equipment that could be used to measure the
         temperature of the water.   (1)

3   Jenna and Saed are investigating the heating power of Bunsen burners. They begin by
    checking whether the burner delivers more heat with the air hole open or with it closed.
    They measure the energy transferred from the Bunsen burner to some water, by finding out the time
    taken for the water to boil.

    a) i)   Name the **independent variable** in this experiment.   (1)

        ii) Name the **dependent variable** in this experiment.   (1)

    b) Other variables should be **controlled** to make this a fair test.

        State which of the following are the **three** most important choices: (3)

        ● the position of the Bunsen burner below the beaker
        ● the thermometer that was used
        ● the volume of water in the beaker
        ● the science lab in which they were working
        ● the time of day
        ● the position of the gas tap (i.e. how much gas flows).

4   This question involves identification of gases. Use the information provided to copy and
    complete the table below. Choose from the following gases: (3)

oxygen   **carbon dioxide**

hydrogen   **sulfur dioxide**

Effect on limewater	pH with universal indicator	Effect on a burning splint	Gas
None	4	Puts it out	
None	7	Goes 'pop'	
Turns it cloudy	6	Puts it out	
None	7	Burns more brightly	

5   Suha is interested in how people measured time
    in the past.

    She makes two candles and draws lines on them.

    a) i) Name the apparatus that Suha uses to measure
            the volume of wax she uses in making the
            candles.   (1)

        ii) Suggest what Suha uses to measure the
            distance between the lines.   (1)

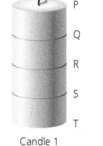

Candle 1   Candle 2

    b) Suha's idea is to time how long it takes for the candles to burn.

        She burns candle 1 first, and presents her results in this table.

Section that was burned	Time taken to burn, in minutes
P to Q	20
Q to R	20
R to S	
S to T	20

        i) Suha draws a bar chart of her values. The values are
            shown on the chart on the right.

            Using a bar chart like the one shown on the right,
            add the missing value.   (1)

        ii) Suha then burns candle 2.

            Draw another bar chart to show roughly how long
            you think it takes for candle 2 to burn between the
            sections.   (1)

    c) Suha thinks that the candles could be used to measure
        time throughout the country. Suggest **three** features of
        the candles that would have to be kept constant if the
        candles are going to be reliable timekeepers.   (3)

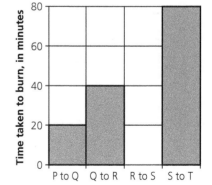

6   Neel used this apparatus to find out which substances are released when ethanol is burned.

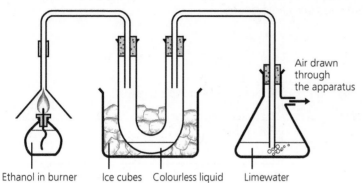

Ethanol in burner   Ice cubes   Colourless liquid   Limewater

a)   i)   Explain why he added ice cubes around the U-tube.   (1)

     ii)   One of the gases turned the limewater milky.

           State the name of this gas.   (1)

     iii) Suggest which test Neel should carry out on the colourless liquid in the U-tube.

           Describe a positive result and suggest what the result would tell him.   (2)

b)   Ethylene glycol (IUPAC) is another alcohol, and it is sometimes used in
     antifreeze. It can be added to the contents of a car radiator to prevent the
     water freezing as the temperature falls.

     i)   There are two hazard warning symbols on the label of the antifreeze
           container.

           State **two** precautions you would take if you were using this
           antifreeze.   (2)

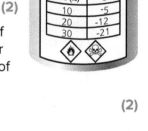

     ii)   The label on the container also provides information about the effect of
           the antifreeze on the freezing point of water. Neel had filled his radiator
           with 2 litres of solution containing 200 cm³ of antifreeze and 1800 cm³ of
           water. During the night, the temperature fell to −18 °C.

           Describe what would happen to Neel's radiator. Explain your answer.   (2)

7   a)   Match the following hazard symbols with their descriptions.
          There are more descriptions than symbols.   (5)

A		Oxidising	
B		Highly flammable	
C		Toxic	
D		Harmful	
E		Corrosive	
		Irritant to eyes or skin	

b) Chemicals can have more than one hazard warning.
State which hazards are identified on this petrol container. (3)

8 Look at these diagrams.

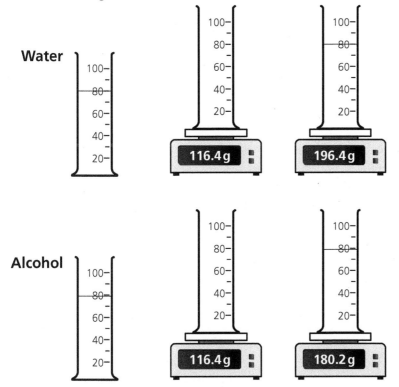

a) Calculate the density of water and of alcohol. Show your working. (3)

b) Oil is less dense than water. If a mixture of oil and water is allowed to stand, the oil will float to the top.
Name a piece of apparatus that will allow you to remove a small volume of oil. (1)

1 Cells and organisation

1 Choose the option which best completes each of the following.

a) Respiration is _____ (1)

producing offspring	**taking in nutrients**
transferring energy	**responding to stimuli**

b) The human male gamete is a _____ (1)

liver cell	**sperm cell**
pollen grain	**nerve cell**

c) Reproduction is _____ (1)

producing offspring	**taking in nutrients**
releasing energy	**increasing in size**

d) Nutrition is _____ (1)

producing offspring	**taking in nutrients**
releasing energy	**responding to stimuli**

e) The basic unit of life is _____ (1)

a molecule	**a tissue**
an organ	**a cell**

f) A collection of cells with the same function is _____ (1)

an organism	**a tissue**
an organ	**a system**

g) The structure inside the cell that carries out aerobic respiration is the

_____ (1)

nucleus	**mitochondrion**
cytoplasm	**chloroplast**

h) A useful stain for observing the nucleus in a cell is _____ (1)

iodine solution	**universal indicator solution**
Benedict's solution	**methylene blue**

2 The human body contains several different systems. The systems are made up of organs working together so that the body is at its most efficient.

This is a list of some of the organs of the human body:

teeth   stomach   testes   rib   heart   lungs

Copy and complete the table below to match the organs listed to the system they belong to. (5)

System	Organs in this system
Digestive	
Circulatory	
Reproductive	
Breathing	
Skeletal	

3 The diagrams below show six cells.

One of these cells transports oxygen in the blood. This cell does not contain a nucleus.

A

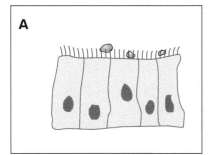

B

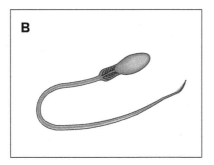

C

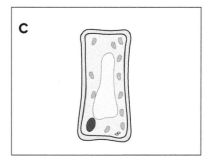

D

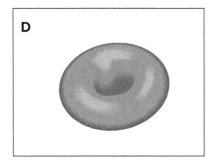

E

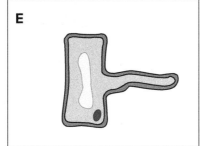

F

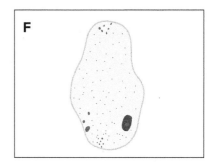

a) i)   Give the letter of the cell that transports oxygen in the blood.   (1)

   ii)   State the function of the nucleus in most cells.   (1)

b) Give the letter of the cell that carries genetic information from father to offspring.   (1)

c) Give the letters of **two** plant cells.   (2)

d) Give the letter of the cell with a surface extended for the uptake of water and minerals.   (1)

4 The diagram shows a plant cell.

   a) i)   This cell is from the leaf of a sycamore tree.

           Name the part that is present in this cell but would
           not be present in a root cell from a sycamore tree.   (1)

      ii)   Explain why the part you have chosen is not present
            in a root cell.   (1)

   b) The parts labelled in this diagram have different functions.

      Copy and complete the table below to link each part
      to its correct function.   (5)

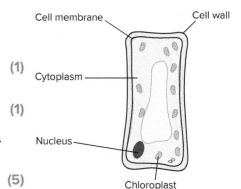

Part of cell	Function
	Helps to give the cell a definite shape
	Controls the entry and exit of substances
	Contains the genetic material that controls the cell's activities
	Many chemical reactions take place here
	Absorbs energy from the Sun for photosynthesis

5   a)   One function of cells lining the human trachea (windpipe) is to

_____   (1)

sweep away particles of dust

release waste carbon dioxide

secrete mucus to trap microorganisms and particles of dust

absorb oxygen for respiration

b)   A cell with a nucleus, a cell wall, no chloroplasts and a large surface
area for absorption is a _____   (1)

red blood cell   **leaf cell**

sperm cell   **root hair cell**

6   a)   The diagram below shows a plant cell and an animal cell.

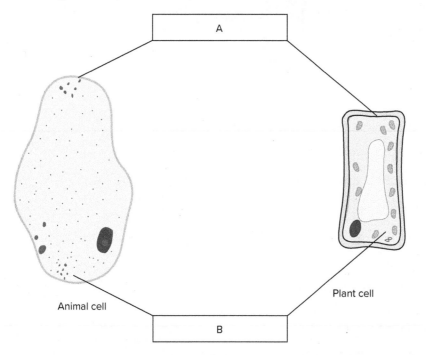

i)   Give the names of **two** parts that are present in plant cells but not in animal cells.   (2)

ii)   Give the function of **one** of the parts you have named and say
why it is important in the life of the plant.   (2)

iii) The letters, A and B, and their guidelines show two parts that are present in both plant and
animal cells.

Identify the two parts and state the function of each of them.   (4)

b) i) Cells can become **specialised**.

Explain what this word means. (1)

ii) Tissues carry out their functions because of the specialised cells they contain.

Copy the words in the boxes below and then link together the cells, their special functions and the biological process they are involved in.

The first cell type has been done as an example. (4)

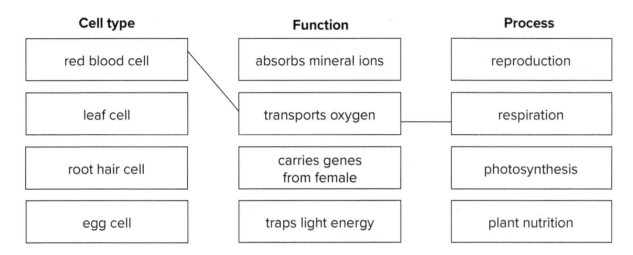

Cell type	Function	Process
red blood cell	absorbs mineral ions	reproduction
leaf cell	transports oxygen	respiration
root hair cell	carries genes from female	photosynthesis
egg cell	traps light energy	plant nutrition

7 The diagram shows a single-celled organism called Chlamydomonas. This organism is able to swim about in the small pools of water where it lives.

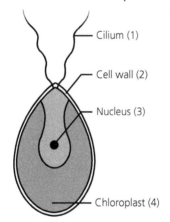

Cilium (1)

Cell wall (2)

Nucleus (3)

Chloroplast (4)

a) In this table, choose which set of numbers (in row A, B, C or D) correctly relates functions of cell parts to the structures labelled in the diagram. (2)

	Function			
	Protection against bursting	Photosynthesis	Movement	Control of cell activities
A	2	4	1	3
B	1	3	2	4
C	4	2	1	3
D	2	4	3	1

b) Name **three** structures in the Chlamydomonas cell that would not be found in a sperm cell. (3)

2 Nutrition and digestion

1 Choose the option which best completes each of the following.

   a) The most important food for muscle growth and repair is _____ (1)

fat	carbohydrate
calcium	protein

   b) Brussels sprouts are a good source of _____ (1)

fat	vitamin D
sugar	fibre

   c) Iodine solution is a stain used to detect _____ (1)

protein	sugar
fat	starch

   d) Calcium is essential in a healthy diet to _____ (1)

prevent scurvy	help develop strong bones
supply energy	help digestion

   e) Egg whites are a good source of _____ (1)

sugar	protein
fat	starch

   f) The most important teeth for biting off pieces of an apple are the _____ (1)

molars	canines
pre-molars	incisors

   g) A positive result in a Benedict's test shows the presence of _____ (1)

starch	an acid
sugar	water

   h) The teeth used by a tiger to kill its prey are the _____ (1)

incisors	premolars
molars	canines

2 a) A poor diet can lead to bad health. Draw lines to match up each fact about the diet to the harm it may cause. (3)

Fact about diet	Harm diet may cause
too much salt	constipation
too little iron	high blood pressure
too much fat	slow growth of muscles
not enough fibre	cannot carry enough oxygen in blood
too little protein	heart disease

b) A balanced, healthy diet should help to prevent this harm. Link each of these components of a healthy diet to its function in a healthy body. (3)

Component of diet	Function in a healthy body
sugar	required for development of bones and teeth
calcium	an important part of the process of digesting foods
vitamin C	the main source of energy for working cells
water	provides a supply of sugar
starch	prevents scurvy

3 The table below provides information about five different foods.

Food	Energy content / kJ per 100 g	Nutrients in 100 g			
		Carbohydrate / g	Fat / g	Protein / g	Calcium / mg
Yoghurt	280	5.0	4.0	3.0	120
Cheese	1700	0.2	35.2	24.0	710
Pear	380	26.1	0.1	0.8	6
Brown rice	900	42.7	1.8	9.5	64
Butter	3118	0	82.1	0.4	17

a) i) State which of the nutrients provides most of the energy in the brown rice. (1)

ii) State which of the four nutrients provides insulation against cold. (1)

iii) State which of the foods would be most useful to a weightlifter needing to build their muscles. (1)

b) i) Calculate the total amount of the three nutrients – fat, protein and carbohydrate – in yoghurt. (1)

ii) State what makes up most of the rest of the 100 g of yoghurt. (1)

c) i) A teenage boy needs about 9000 kJ of energy every day. Calculate how much brown rice he would need to eat to obtain this amount of energy. (1)

ii) The boy also needs about 55 g of protein every day. Explain whether this same amount of rice provides all of his protein requirements. (1)

d) The table below shows the recommended daily amount (RDA) of calcium for a female at different times in her life.

Stage of life cycle	RDA of calcium / mg
Baby aged 3 months	450
12-year-old girl	900
21-year-old, non-pregnant woman	550
Pregnant woman	1200
Breast-feeding woman	

i) Suggest what would be the RDA of calcium for a breast-feeding woman. Explain your answer. **(1)**

ii) Explain why the 12-year-old girl has a higher RDA of calcium than the 21-year-old woman. **(1)**

iii) The 21-year-old woman needs only about three-quarters of the protein as the 12-year-old girl. Explain why we need protein in our diet. **(1)**

4 Teeth develop properly only if the diet contains plenty of _____ **(1)**

starch   **vitamin C**

calcium   **iron**

5 Scientists are sure that a healthy diet reduces the risk of disease. They often recommend certain foods to improve health, for example, foods low in cholesterol are known to reduce the risk of heart disease. One of these recommended foods is mycoprotein, an artificial 'meat' made from harmless fungi.

The bar chart shows the levels of different nutrients in beef and in mycoprotein.

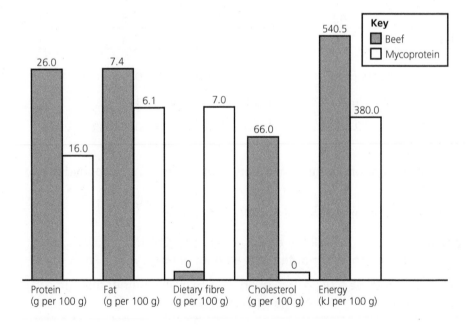

a) i)   Explain **three** reasons why mycoprotein is healthier than beef.   (3)

    ii) Give **two** reasons why it might be better to eat beef than mycoprotein.   (2)

b) Scientists also recommend that we eat more fibre in our diet. They have compared the intake of fibre with the chance of developing colon cancer (the colon is part of the large intestine). This scatter graph shows their results.

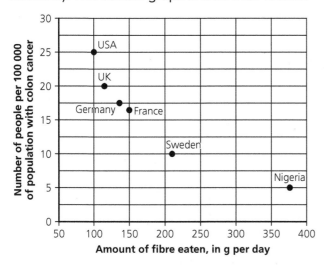

i)   Identify which country had the largest proportion of people with colon cancer.   (1)

ii) Calculate how much more likely it is for a person in the UK to have colon cancer than one from Nigeria. Explain how you reached your answer.   (2)

iii) State which **two** of the following foods are good sources of fibre.   (1)

    wholemeal bread   **chocolate**

    cheese   **apples**

    eggs   **pizza**

6 Sally is investigating the pH of milk left in a sealed container for five days. She obtains the following results:

Time, in days	pH of sealed milk sample
0	7.1
1	6.4
2	5.9
3	5.2
4	4.5
5	4.1

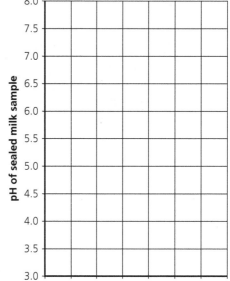

a) Plot these results on a graph grid like the one shown on the right.   (3)

b) Sally believes that bacteria are causing this change in pH. Suggest what the bacteria are releasing to bring about this change.   (1)

c) Use the graph to predict the likely pH after 7 days.   (1)

d) Sally decides to try to find out whether heating the milk would stop this pH change. Describe in detail how she could do this. Use the terms **independent variable**, **dependent variable** and **controlled variable** in your answer. (4)

e) People use this information about the effect of bacteria on milk in many ways. Name **one** food that we use that is made by letting bacteria turn milk sour. (1)

f) If we wish to prevent milk from going sour, there are many things we can do to it. Choose **two** treatments from this list that would help to prevent milk from going sour. (1)

- Keep the milk at a low temperature in a refrigerator.

- Mix the milk with fruit juice.

- Dry the milk and store it as granules.

- Always shake the milk before pouring it.

For **one** of your choices, explain why the treatment will stop the milk from going sour. (1)

7 Ehsaan carried out an investigation into the way the human body digests and absorbs starch. He used amylase solution to digest the starch, and special bags made of semi-permeable membrane to act as a model for the walls of the intestine. The experiment was set up as shown in the diagram below.

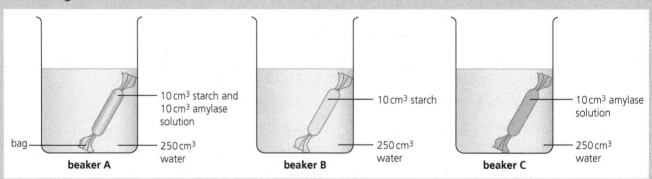

The water in the beakers was kept at 37°C and after 20 minutes, Ehsaan tested the contents of each beaker and bag for starch and sugar.

a) i) Name the reagent (solution) used to test for starch. (1)

ii) Name the reagent (solution) used to test for sugar. (1)

b) Suggest why Ehsaan kept the water temperature at 37°C. (1)

The table shows Ehsaan's results.
+ = a positive result, − = a negative result

Beaker	Test on contents of bag		Test on water from beaker	
	starch	sugar	starch	sugar
Beaker A	+	+	−	+
Beaker B	+	−	−	−
Beaker C	−	−	−	−

c) i) Explain the results for starch in the water from the beaker. (1)

ii) Explain why sugar was found in the bag in beaker A. (1)

d) Suggest the function of beaker C. Choose **one** of the following:

as a fair test   **for accuracy**

as a control   **for reproducibility** (1)

e) The following diagrams can be used to explain what happened in the investigation.

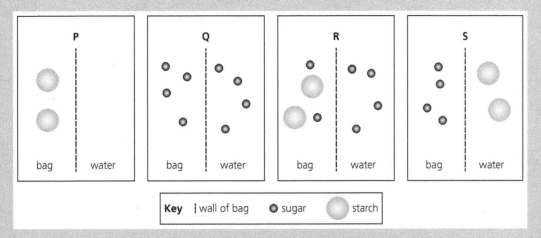

i) State which diagram, P, Q, R or S, represents the results of beaker A. (1)

ii) State which diagram, P, Q, R or S, represents the results of beaker B. (1)

f) Ehsaan made an important assumption when he suggested that this is the way the human body digests starch. Suggest what this assumption was. (1)

3 Respiration, energy and exercise

1 Choose the option which best completes each of the following.

a) Gas exchange in humans takes place in the _____ (1)

   windpipe/trachea   **liver**

   alveoli/air sacs   **mouth**

b) A gas that turns limewater milky is _____ (1)

   oxygen   **carbon dioxide**

   carbon monoxide   **water vapour**

c) The gas required by respiring cells is _____ (1)

   carbon dioxide   **water vapour**

   nitrogen   **oxygen**

d) Cigarette smoking does not cause harm to the _____ (1)

   lungs   **fingernails**

   heart   **unborn baby**

e) The transfer of energy by the oxidation of food is _____ (1)

   digestion   **absorption**

   excretion   **respiration**

f) The normal resting pulse rate in a healthy individual is approximately

   _____ (1)

   40 bpm   **120 bpm**

   70 bpm   **100 bpm**

g) A disease which reduces the efficiency of the lungs is _____ (1)

   cholera   **athlete's foot**

   diarrhoea   **emphysema**

h) A condition in which restricted breathing can be treated by an inhaler is

   _____ (1)

   asthma   **tiredness after exercise**

   lung cancer   **accidental swallowing of a fizzy drink**

2 The diagram shows the rib cage in a human.

   a) The rib cage is able to move during breathing.

      i) Name the type of tissue responsible for this movement. (1)

      ii) State the direction in which the ribs move when we
          breathe in. (1)

   b) The rib cage also plays a part in protecting delicate organs.
      Give the names of **two** organs that the rib cage protects. (2)

Breast bone

Rib

Cartilage

3 David wants to compare the energy value of several foods. He uses the apparatus shown below and measures the rise in temperature caused by the burning food sample.

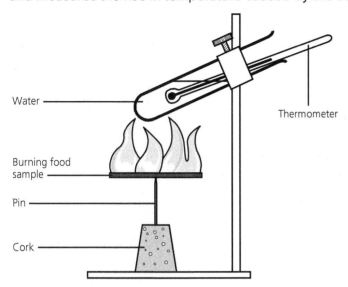

Water

Thermometer

Burning food sample

Pin

Cork

a) i)   What is the **independent variable** in his experiment?   (1)

    ii) What is the **dependent variable** in his experiment?   (1)

    iii) Suggest **two** steps that David should take to make this a fair test.   (2)

    iv) David's teacher suggests collecting all of the class results together before trying to draw conclusions. Explain why this is important.   (2)

b) i)   In a living cell, the food does not burn in this way to transfer energy. Name the process that transfers energy from food stores in living cells.   (1)

    ii) Give **two** reasons why energy is required in the body.   (2)

4 a) When cigarette smoke is bubbled through universal indicator solution, the solution changes

       from _____   (1)

       red to yellow   **blue to green**

       green to red-orange   **green to blue**

   b) Anaerobic respiration in animals produces _____   (1)

       glucose   **some energy and carbon dioxide**

       alcohol and carbon dioxide   **some energy and lactic acid**

5   Anika is a good athlete and wants to find out how training is affecting her breathing. She is able to use a machine that measures the volume of air breathed in and out – the machine allows measurements to be made before and during exercise.

a) The chart shows results obtained during one investigation.

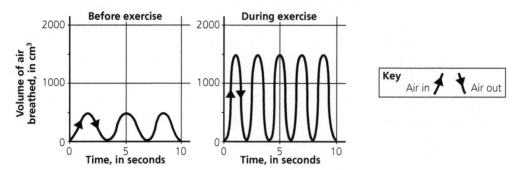

i)   Calculate how much more air Anika breathes in with each breath during exercise (in cm³).   (1)

ii)   The air contains 20% oxygen. Calculate how much more oxygen Anika breathes in per minute. Show your working.   (4)

iii) Copy and complete this word equation to explain why Anika breathes in this extra oxygen during the period of exercise.   (2)

_____ + oxygen → _____ + water + _____

b) Suggest which other organ in Anika's body works faster to help this extra oxygen reach the parts of the body where it is needed.   (1)

c) Sometimes an athlete exercises so hard that not enough oxygen can be obtained. Under these circumstances, respiration becomes anaerobic.

i)   State **two** reasons why this form of respiration is less valuable to the athlete.   (2)

ii)   Anaerobic respiration can also occur in other organisms. Name **one** organism that can carry out anaerobic respiration to make a product useful to humans.   (1)

Name this useful product.   (1)

6   The diagram below shows part of a human breathing system.

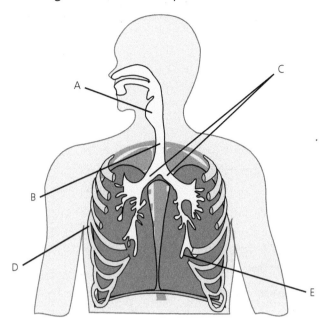

a)   Give the letters that label:

i)   the part that protects the lungs (1)

ii)   the part where gases are exchanged between blood and air. (1)

b)   Carbon monoxide damages the cells lining the windpipe, and tar irritates these cells so that they make more mucus. Suggest **one** result of this for a cigarette smoker. (1)

c)   i)   Name the gas which passes from air to blood in the part you have identified in **a) ii)**. (1)

ii)   Name the gas which passes from blood to air. (1)

iii) Name the process that moves the particles of these gases. (1)

iv) The lining that these gases cross is thin and has a large surface area. Explain why each of these features is important. (2)

7   a)   Some people are worried that if they eat too much fat they will become overweight. In humans, energy is transferred from fatty foods into long-term stores of energy. The energy content of a fatty food can be investigated using the same apparatus as shown in Question 3 on page 23.

4.2 J of energy will raise the temperature of 1 cm³ of water by 1°C. 1 g of fat contains 38 500 J of energy.

Calculate the rise in temperature of 25 cm³ of water if 0.2 g of fat is burned in this way. Show your working and give your answer to the nearest whole number. (3)

b)   In the actual experiment, the temperature rise was much less than expected. The science teacher suggested that this might be due to transfer of energy to the surroundings. Suggest **two** ways in which energy might be lost and so not heat up the water. (2)

4 Reproduction in humans

1 Choose the option which best completes each of the following.

a) The human male gamete is _____ (1)

an ovum	a sperm
an antibody	a zygote

b) On average, a woman ovulates every _____ (1)

3 weeks	28 weeks
28 days	3 months

c) The process when gametes join together is called _____ (1)

ovulation	menstruation
gestation	fertilisation

d) Sperm are produced in the _____ (1)

sperm duct	scrotum
testes	penis

e) The genes from the two parents are carried in the part of the sex cell called the

_____ (1)

cytoplasm	nucleus
membrane	embryo

f) The stage of human development at which a person becomes able to reproduce is called

_____ (1)

adulthood	puberty
gestation	birth

g) Two gametes join together to form _____ (1)

an embryo	a fetus
a zygote	a gamete

h) The part of the body where substances can be exchanged between a pregnant woman
and her developing baby is the _____ (1)

liver	placenta
umbilical cord	amniotic sac

i) The length of time between fertilisation and birth is called _____ (1)

gestation	menstruation
conception	copulation

2   The diagram shows the reproductive system of a male.

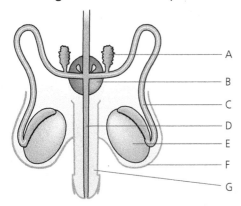

a)   State which of the labelled structures (choose the correct letter in each case):

   i)   produces sperm   (1)

   ii)   carries urine as well as sperm   (1)

   iii) produces a fluid for sperm to swim in.   (1)

b) i)   Identify which structure, shown in the diagram, is cut in a contraceptive operation.   (1)

   ii)   Explain why this operation is a successful form of contraception.   (1)

c)   A chemical called testosterone is produced in a boy's body from adolescence onwards. This chemical causes certain changes in the boy's body. Describe **two** changes caused by testosterone.   (2)

3   This diagram shows a fetus in the uterus just before birth.

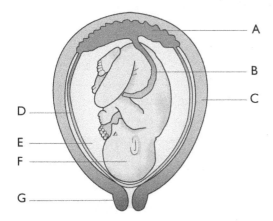

a)   State which letter labels:

   i)   the amniotic sac   (1)

   ii)   the umbilical cord   (1)

   iii) a muscle that can push out the baby at birth.   (1)

b)   State the normal length of pregnancy in humans in months.   (1)

c)   State the function of the amniotic fluid around the fetus.   (1)

d)   The pregnant woman sometimes has cravings for particular foods. Many women like to have milk chocolate, which contains a lot of sugar. Explain in detail how this sugar reaches the developing fetus while the fetus is inside its mother's uterus.   (4)

4   The diagrams below show some cells involved in reproduction.

A

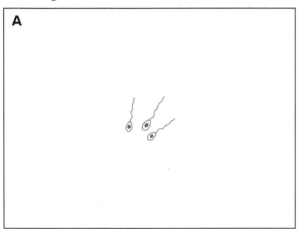

B

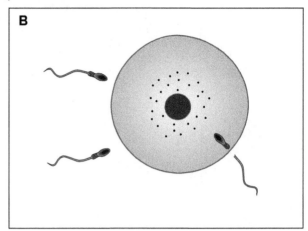

a) i)   State the name of cells A.   (1)

    ii) Give **two** ways in which this cell is adapted to its function.   (2)

b) i)   Name the process shown in B.   (1)

    ii) Name the part of the body where this process takes place.   (1)

c) Copy and complete the following sentences.   (4)

    About six days after process B occurs, a ball of cells called an _____

    becomes embedded in the thickened wall of the _____. This process is called

    _____, and once it has successfully been completed, a new structure called the

    _____ forms, linking the mother to her developing baby.

5   This question is about the menstrual cycle.

a) Copy and complete the following sentences.   (5)

    One of the _____ releases an egg cell (ovum) every _____.
    If there is no fertilisation, or the fertilised egg does not stick to the lining of the womb, then
    _____ occurs. When this process takes place, a woman loses _____.
    This is often called 'having a _____'.

b) This diagram shows the lining of the uterus during the menstrual cycle.

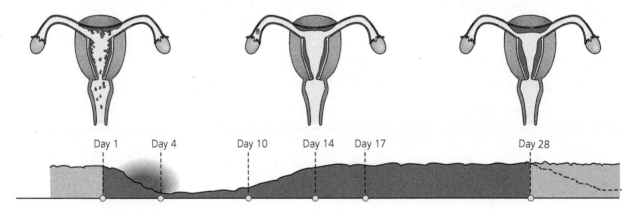

    Between or around which days is the time:

    i)   of ovulation?   (1)

    ii)   when the uterus lining is lost?   (1)

    iii) when fertilisation is most likely?   (1)

6   During pregnancy a woman's body undergoes many changes. One noticeable change is that her body grows as the fetus grows. This table shows the changes in mass of some of her body parts. Answer the questions below.

Body part	Increase in mass, in kg
Uterus	1.0
Breast tissue	0.4
Fat	3.7
Placenta	0.7
Blood	0.4
Amniotic fluid	0.8

a) i)   Use the figures in the table to plot a bar chart on a grid like the one below. Do not include the figure for fat in your bar chart. (3)

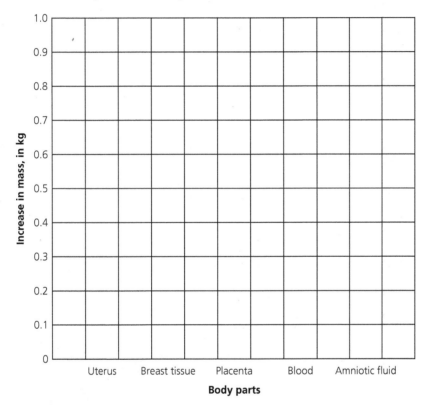

ii)   The pregnant woman also gains about 1.5 kg of water and about 0.5 kg of bone. Calculate the total gain in mass, including the figure for fat, during pregnancy. (1)

iii)   Calculate what proportion of this gain in mass is due to the placenta. Show your working. (2)

b)   Explain why the pregnant woman must make sure that she has plenty of protein and calcium in her diet. (2)

c)   Women who smoke during pregnancy risk harming their babies. Tobacco smoke contains many chemicals, including nicotine and carbon monoxide gas (this gas combines with red blood cells and reduces their capacity to transport oxygen).

Explain how these might cause damage to the developing fetus. (2)

1   Choose the option which best completes each of the following.

    a)   The anther of a flower _____   (1)

        receives pollen   **attracts insects**

        produces pollen   **holds up the stigma**

    b)   The male gamete in a pollen grain does not have _____   (1)

        a membrane   **cytoplasm**

        a nucleus   **chloroplasts**

    c)   A reagent that can be used to test for starch in a seed is _____   (1)

        universal indicator   **limewater**

        iodine solution   **methylene blue**

    d)   A plant cell wall is made of _____   (1)

        protein   **fat**

        starch   **cellulose**

    e)   The part of a flower that receives pollen from a visiting insect is the _____   (1)

        stigma   **anther**

        petal   **sepal**

2   a)   This is a diagram of a single grass flower.

        Match the letters A to G with labels chosen from this list: (7)

        anther

        filament

        ovary

        pollen grains

        stamen

        stigma

        style

    b)   Describe and explain **two** features of this flower that make pollination more likely to be successful.   (2)

3 **a)** The diagram below shows an experiment on the germination of pea seeds.

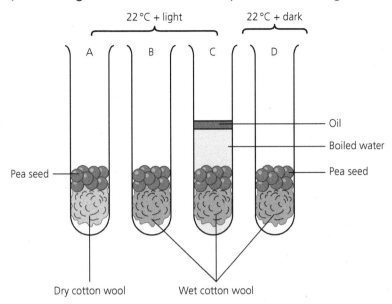

Copy and complete this table. **(4)**

Tube	Would seeds germinate? (write YES or NO)
A	
B	
C	
D	

b) Sara and Yutong measured the dry mass of some germinating barley seedlings for the first 35 days after sowing. Their results are shown in this table.

Time after sowing / days	0	7	14	21	28	35
Dry mass / g	4.0	2.8	2.8	4.4	6.8	8.6

    i) Plot these results in a line graph. Use the graph grid below. **(2)**

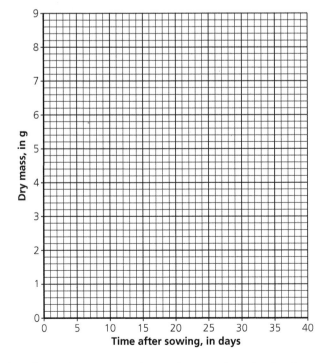

    ii) State how many days after sowing the barley seedlings regained their original mass. **(1)**

    iii) Explain why the dry mass falls in the first stages of germination. **(1)**

4 Choose the option which best completes each of the following.

   a) The roots of a small plant grow _____ (1)

   towards light and against gravity   **with gravity and away from light**

   towards light and with gravity   **towards both light and water**

   b) In the plant life cycle, insects are most important for _____ (1)

   dispersal   **pollination**

   fertilisation   **photosynthesis**

   c) The part of the seed that protects it from bacteria and fungi is the _____ (1)

   carbohydrate store   **micropyle**

   seed leaf   **testa/seed coat**

   d) For germination, a seed requires _____ (1)

   oxygen, water and light

   water, oxygen and a suitable temperature

   water, carbon dioxide and a suitable temperature oxygen

   light and a suitable temperature

5 This diagram shows a section through an insect-pollinated flower.

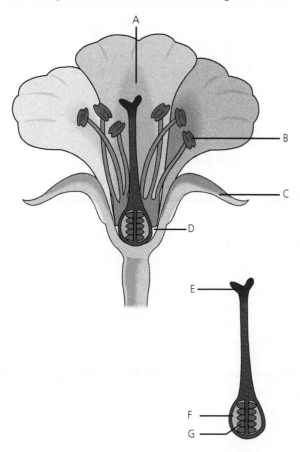

   a) Use the letter labels to identify which part of the flower:

      i)   makes the male gametes (1)

      ii) attracts insects (1)

      iii) will eventually become a fruit (1)

      iv) receives pollen. (1)

b) This paragraph is about the life cycle of plants. Copy and complete it by filling in the missing words, chosen from this list: (5)

dispersal   **fertilisation**   **pollination**   **germinates**   **reproduction**   **grows**

A young plant develops when a seed _____ and matures until it produces

a flower for _____. Male gametes are transferred to the plant during the process of

_____ and join with female gametes during _____. Eventually fruits

are formed and are separated from the parent plant during the process of _____.

c) Jack noticed that birds often eat seeds and his teacher explained that the seeds contained food stores, such as starch. Describe in detail how Jack could test that seeds contain starch. Include any control he would need to use and give **one** safety precaution he should take. (4)

6 Some lily plants produce heat to raise the temperature of their flowers. The heat attracts insects to the flower and is transferred during the process of aerobic respiration.

a) State **one** benefit to the plant of attracting insects. (1)

b) Aerobic respiration uses oxygen. A scientist was interested in the relationship between the uptake of oxygen and the temperature of the flowers. He took 20 lily flowers and measured the temperature and the rate of
oxygen uptake for each one of them. The results are shown in the graph below.

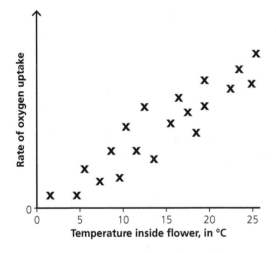

i) Describe the relationship between the temperature inside the flower and the rate of oxygen uptake. (1)

ii) Explain the reasons for the relationship you have described. (2)

c) i) Name the chemical store of energy that the lily plant uses for aerobic respiration. (1)

ii) Explain how the lily obtains this chemical store energy. (2)

6 Healthy living

1 Choose the option which best completes each of the following.

a) A disease that can be passed on to another, unrelated person is

_____ (1)

infectious	**fatal**
inherited	**caused by lifestyle**

b) The benefits of exercise do not include _____ (1)

increased stamina	**greater strength**
athlete's foot	**a stronger heart**

c) When the body cannot function without a drug, the person is said to be

_____ (1)

acclimatised	**activated**
adapted	**addicted**

d) An example of an infectious disease is _____ (1)

lung cancer	**heart disease**
COVID-19	**depression**

e) Excessive use of alcohol can cause _____ (1)

tuberculosis	**a cold**
AIDS	**liver damage**

f) A medicinal compound that can reduce the growth of bacteria inside the body is

_____ (1)

an antiseptic	**an antibody**
aspirin	**an antibiotic**

g) Fibre is needed in the diet to reduce the risk of _____ (1)

constipation	**liver damage**
brain tumours	**weakened bones**

2 The graph below shows how the number of cigarettes smoked each day affects the percentage of deaths from heart disease in the male UK population, at different ages.

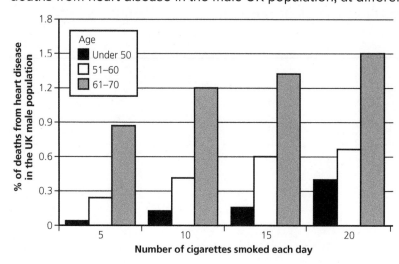

a) Write **two** conclusions about the effect of smoking on heart disease in males. (2)

b) This diagram shows the effect of smoking on the arteries of the heart.

i) Explain how this could cause damage to the heart. (2)

ii) Smoking also causes a rise in blood pressure. Explain how this could affect the health of a smoker. (1)

Healthy artery   Artery of smoker

c) Anna smokes every day. This graph shows the amount of nicotine in her blood after smoking a cigarette. She feels the need for a cigarette once the nicotine level falls below the 'demand threshold'.

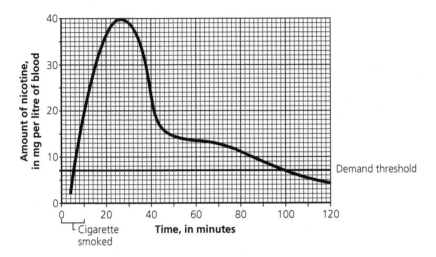

i) State how often Anna needs to smoke to keep the nicotine level above the threshold. (1)

ii) Suggest why Anna feels anxious and bad-tempered when she wakes up in the morning. (1)

iii) Smoking 20 cigarettes per day doubles the demand threshold. State what the new demand threshold would be. (1)

iv) Suggest what effect this will have on the length of time that Anna can wait between cigarettes without becoming stressed. (1)

v) Suggest **one** other way that Anna can satisfy her craving without smoking a cigarette. (1)

3 Drinking alcohol causes changes in the way the body works.

a) Copy the words in the boxes below and then draw lines to link the changes to the effects they have on a drinker's actions and health. (3)

Change	Effect on actions and health
nerve impulses travel more slowly	poor judgement of distance
blood vessels close to the skin open up	long-term liver damage
senses work less well	person looks red-faced
liver cells try to remove alcohol from blood	reactions are slowed

b) A pregnant woman can pass any alcohol she drinks to the fetus.

    i) Describe in detail how the alcohol would reach the fetus. (3)

    ii) If the woman also smokes, she may harm her unborn baby even more. Suggest how chemicals in smoke might affect the health of the unborn baby. (2)

4 Choose the option which best completes each of the following.

  a) A bacterial cell carries its genetic information in its _____ (1)

    membrane   **DNA**

    nucleus   **cell wall**

  b) One disease caused by a bacterium is _____ (1)

    influenza   **athlete's foot**

    typhoid   **AIDS**

5 This diagram shows a simple virus.

Viruses reproduce inside living cells and often cause disease.

  a) Name **one** disease caused by a virus. (1)

  b) Complete the following paragraph about disease. Use words from this list: (3)

    antibiotic   **transfusion**   **painkiller**   **antibody**   **vaccination**

    Diseases caused by viruses can be prevented by a _____. Following this process, the number of _____ molecules in the blood increases. An illness caused by a virus cannot be successfully treated with an _____.

6 It is possible to transplant organs into patients suffering from some diseases, such as heart failure. A patient who receives a transplanted heart may reject the new heart because the immune system recognises the new organ as 'foreign' to the body.

  a) The patient can take drugs to slow down their immune system. Suggest **one disadvantage** to the patient of taking these drugs. (1)

  b) The patient may also have to take antibiotics following transplant surgery. Some of the organisms that might cause infection are resistant to these antibiotics. MRSA is the name given to an organism that is resistant to a commonly used antibiotic.

    i) Suggest **two** ways in which MRSA could get into the body of a patient. (2)

    ii) The nurses looking after the patient are very careful to reduce the risk of infections: Explain how the following precautions help to reduce the risk of infection:

       • wearing a surgical mask when in the patient's room (1)

       • keeping the windows of the ward closed (1)

c) A transplant is less likely to be needed if the heart is healthy. Some aspects of our lifestyle can help to protect the heart against disease. Which items from the list below would benefit the heart? (3)

- a diet low in animal fats
- regular exercise
- a diet low in vitamin C
- a diet with twice the normal level of protein
- a lifestyle with little stress

d) The community is also responsible for fighting disease. State which **one** of the following is not a community responsibility. (1)

- providing safe drinking water
- removal of sewage
- making sure that children have new clothes
- checking on hygienic food preparation

7 Tolani and Fumi eat some soft ice cream on school Sports Day. There is food-poisoning bacteria in the ice cream and they become quite ill the next day. The doctor gives them antibiotics and tells the sisters to take them for eight days. This graph shows the change in the number of bacteria in the body if antibiotics are taken.

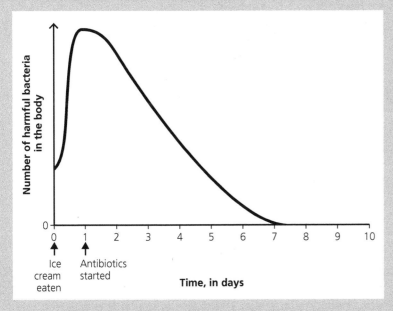

a) i) Explain why the sisters did not become unwell until the day after eating the ice cream. (1)

ii) Tolani feels much better after taking the antibiotics for eight days. Explain why. (1)

iii) Fumi is the older sister and thinks she knows better! After four days, she feels fine and so stops taking the antibiotics. Two days later, she is really ill again. Explain why this happens. (1)

b) i) Food poisoning can make a person vomit and have diarrhoea. This makes them lose water. Give **one** important function of water in the body. (1)

ii) Vomiting can bring acid from the stomach into the mouth. Explain how this could harm the teeth. (1)

c) It is possible to give a vaccine against some types of bacteria that cause diarrhoea. Explain how a vaccine helps to control infection by bacteria. (2)

8   Ferdinand Magellan led the first voyage around the world. He sailed from Spain in 1519, hoping to find a new route to the Spice Islands.

There were five ships in his small convoy, with 237 men. Each ship carried a supply of basic foods including flour, cheese, dry biscuits, oil, meat and vegetables. The ships arrived at the Spice Islands after a difficult voyage of 20 months, but only one was able to return to Spain. The sailors on this ship became very ill before they reached Spain again – they had sores that would not heal and their teeth fell out of their bleeding gums.
One sailor, named Elcarno, ate a spoonful of fruit jam every day and he did not develop any of these symptoms.

a)   The mixture of foods taken by Magellan's ships did not provide a healthy diet. Explain what is meant by a healthy diet.   (2)

b)   The sailors developed a deficiency disease called scurvy on the return voyage to Spain.

   i)   Explain what is meant by a deficiency disease.   (1)

   ii)   Suggest the cause of scurvy.   (1)

   iii) Describe **one** symptom of scurvy.   (1)

   iv)   Suggest why the sailor Elcarno did not develop this deficiency disease.   (1)

   v)   Name **one** other food that would reduce the risk of developing scurvy.   (1)

7 Photosynthesis

1 Choose the option which best completes each of the following.

a) During photosynthesis a leaf uses _____ (1)

oxygen   **starch**

carbon dioxide   **protein**

b) Root hair cells have a large surface area to _____ (1)

photosynthesise more efficiently   **absorb minerals and water**

make contact with other roots   **store excess carbohydrate**

c) An animal cell does not have _____ (1)

a membrane   **cytoplasm**

a nucleus   **chloroplasts**

d) A plant cell differs from an animal cell because the plant cell has _____ (1)

a nucleus   **cytoplasm**

a cell membrane   **a cell wall**

e) Energy transferred to the plant in photosynthesis comes from _____ (1)

the soil   **decomposition of carbohydrates**

the Sun   **heat in rainwater**

f) The roots of a plant absorb _____ (1)

water and minerals   **water and carbon dioxide**

carbon dioxide and minerals **nitrates and light**

g) The part of the plant cell that is a store for energy transferred as light is the _____ (1)

cell wall   **cell membrane**

chloroplast   **nucleus**

h) Within the plant, the first store of energy during photosynthesis is _____. (1)

water   **starch**

chlorophyll   **oxygen**

2 Andy carries out an experiment to investigate whether carbon dioxide is needed for photosynthesis. The experiment is set up as shown in the diagram below:

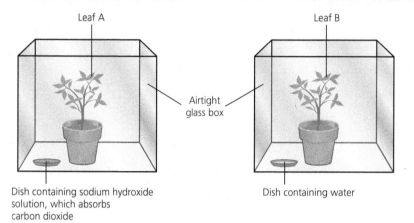

Leaf A

Leaf B

Airtight glass box

Dish containing sodium hydroxide solution, which absorbs carbon dioxide

Dish containing water

To find out whether photosynthesis has taken place, Andy tests leaves for the presence of starch. Starch is made from glucose.

a) i) Write out the word equation for the process of photosynthesis. (2)

ii) Name the chemical that is used to test for starch. (1)

iii) Describe a positive result of such a test. (1)

b) State which one of the leaves you would expect to show the presence of starch. (1)

Explain your answer. (2)

c) i) Describe what is unusual about leaf C in this diagram. (1)

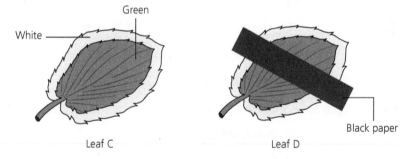

Green

White

Black paper

Leaf C

Leaf D

ii) Describe how you would remove all of the starch from these leaves. (1)

iii) Both of these leaves were destarched and then exposed to bright light for 6 hours. Draw a diagram to show what the leaves would look like if they were now tested for starch. (2)

3 Billy buys a potted plant for his grandma's birthday. She tries to look after it, but after a while she notices that the leaves are turning yellow.

a) Name the green pigment in plants. (1)

b) A potted plant is more likely to suffer from a mineral shortage than a plant growing in the garden. Explain why. (1)

c) Billy wants to buy some houseplant fertiliser for his grandma's plant. His science teacher says that he could mix his own and suggests the following substances:

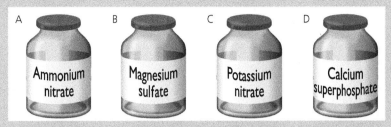

i) Give the letter of one substance he needs to include to make sure that his fertiliser contains each of the following minerals: (2)
- potassium _____
- phosphate _____
- nitrate _____
- magnesium _____

ii) Sometimes excess fertilisers can be washed into lakes and rivers. If this happens, the oxygen content of the water can fall. This can prevent organisms from carrying out aerobic respiration. Give **two** ways in which anaerobic respiration is less useful than aerobic respiration. (2)

iii) Some of the leaves fall off grandma's plant and lay on the soil beneath it. They slowly break down into simpler molecules. Name **one** type of organism that carries out this breakdown. (1)

4 Choose the option which best completes each of the following.

a) A process that can raise carbon dioxide levels in the environment is
_____ (1)

denitrification   **photosynthesis**

combustion   **leaching**

b) Bacteria and fungi are useful in the environment because they _____ (1)

work as producers   **act as carnivores**

break down waste materials   **are photosynthesisers**

c) While it is dark, a plant _____ (1)

respires and photosynthesises

respires only, using oxygen

photosynthesises only, using carbon dioxide

respires only, using carbon dioxide

5   Amna uses the apparatus shown below to measure how the concentration of dissolved carbon dioxide affects the rate of photosynthesis.

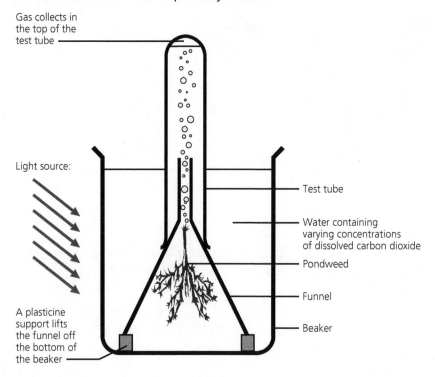

She obtains the following results:

Concentration of dissolved carbon dioxide as %	Rate of photosynthesis, in number of bubbles released per minute
0.05	7
0.10	13
0.15	20
0.20	26
0.25	30
0.30	31
0.50	31

a) She plots the results on a grid to obtain the graph below.

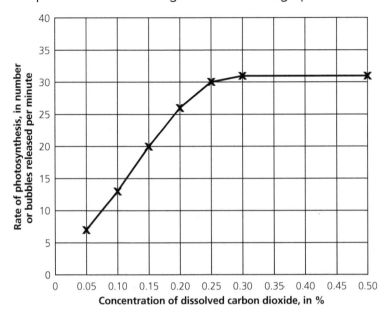

i) Use the graph to find at what concentration of dissolved carbon dioxide the plant produced 24 bubbles per minute. (1)

ii) Explain how the graph tells you how a greenhouse owner could grow their plants most efficiently. (2)

b) When Amna carries out this investigation, she wants it to be a fair test. Name **three** factors she should control for this to be true. (3)

c) Amna thinks that the gas given off by the plant is oxygen.

i) Describe how she can test if this is true. (1)

ii) Amna's teacher says that the number of bubbles released is not an accurate way of measuring the rate of photosynthesis. Suggest how Amna can measure the volume of gas more accurately. (1)

iii) Suggest how Amna can make her results more reliable. (1)

6   Hydrogencarbonate indicator solution changes colour according to changes in pH.

pH	Colour of indicator
Neutral or very slightly acidic	Red
Acidic	Yellow
Alkaline	Purple

Five test tubes are set up as shown in the diagram below. Red hydrogencarbonate indicator solution is added to each of the tubes. The tubes are left on a sunny window ledge for 3 hours.

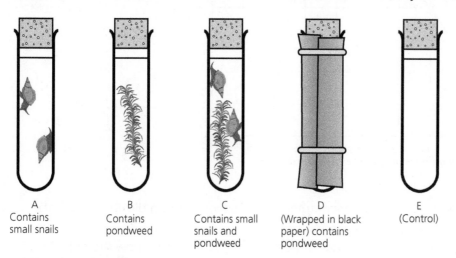

A	B	C	D	E
Contains small snails	Contains pondweed	Contains small snails and pondweed	(Wrapped in black paper) contains pondweed	(Control)

a) i)   Name the acidic gas, produced by living organisms, which is likely to affect the acidity of the indicator solution. (1)

    ii) After 3 hours, describe the likely colour of the indicator solution in test tube A. (1)

    iii) Explain your answer. (1)

b) i)   Describe the colour of the indicator solution in tube B. (1)

    ii) Explain your answer. (1)

    iii) The colour of the indicator solution does not change in tube C. Explain why. (1)

c) Explain the purpose of the control (tube E). (1)

8 Interdependence of organisms in an ecosystem

1. Choose the option which best completes each of the following.

   a) Fungi can _____ (1)

   | photosynthesise | attack and eat small animals |
   | break down waste materials | replace plants in the environment |

   b) Plants increase the biomass in the environment through the process of

   _____ (1)

   | absorption | respiration |
   | seed production | photosynthesis |

   c) The transfer of energy between energy stores in different living organisms can be

   drawn out as _____ (1)

   | an example of respiration | a food chain |
   | a pyramid of numbers | photosynthesis |

   d) A chemical used to detect a product of photosynthesis is _____ (1)

   | iodine solution | hydrogencarbonate indicator solution |
   | litmus solution | oxygen gas |

   e) A simple method of measuring the size of a population uses a _____ (1)

   | ruler | measuring cylinder |
   | quadrat | set square |

   f) In a simple food chain, a predator would be _____ (1)

   | a herbivore | a fungus |
   | a carnivore | the Sun |

   g) The producer in an ecosystem is always a _____ (1)

   | large animal | fungus |
   | green plant | small animal |

2. Copy the words in the boxes below and then draw lines to match up each term with the best description. There are more descriptions than terms. (4)

Ecological term	Description
Reproduction	All the members of the same species living in one area
Competition	Managing the environment for the benefit of wildlife
Population	Reduces the number of organisms in a habitat
Conservation	Leads to an increase in population
	Two or more organisms trying to obtain the same thing from their environment

3   The diagram shows a food web in the sea close to Antarctica.

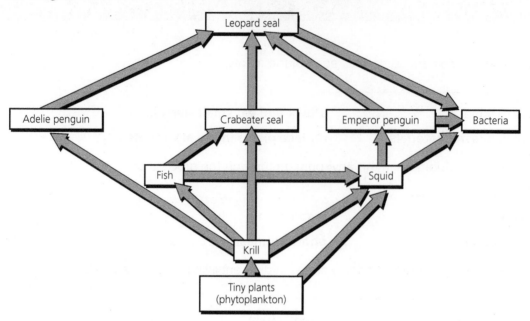

a)   i)   Identify an example of each of the following: (4)
         ● a herbivore
         ● a producer
         ● a carnivore
         ● an organism that breaks down waste materials

     ii) Draw out a food chain of five organisms selected from this food web.   (2)

b)   Emperor penguins feed on squid. Squid swim very quickly and have a slippery skin.

         Suggest **two** ways in which the emperor penguin is well adapted to catching its prey.   (2)

4   Choose the option which best completes each of the following.

a)   The final size of a population is not affected by _____   (1)

     the method used to count the organisms

     competition for food

     the number of disease-causing organisms

     the number of predators

b)   The top carnivore in a habitat is always _____   (1)

     a bird   **a fox**

     very small   **an animal**

c) Each of the following is an example of pollution except for _____ (1)

excess chemicals flowing into rivers

poisonous chemicals being sprayed onto crops

woodland being cut down

sulfur dioxide being released from car engines

d) A habitat does not provide _____ (1)

food   **breeding sites**

predators   **shelter**

5 The graph below shows how the population of wild trout in a lake changed over a period of time.

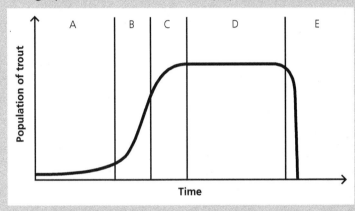

a) i) What does the section labelled D tell you about the birth rate and the death rate of
        the trout during this time period?   (1)

    ii) Explain how you know this.   (1)

b) i) State which part of the curve shows when the fish began to compete with each other
        for food.   (1)

    ii) Explain how you know this.   (1)

    iii) Suggest **one** other factor that might be affecting the population curve at this point. (1)

c) A fish farmer decides to grow trout in enormous nets in a Scottish loch. He provides the
    food for the trout and then catches them for sale.

    i) He wants the trout to grow quickly. Suggest which nutrient the trout food should
        contain to make sure that this happens.   (1)

    ii) The trout food is expensive and the fish farmer wants to make a profit. A population
        of farmed trout will follow the same curve as the population of wild trout shown in
        the graph. State which section of the population curve represents the best time for him
        to catch the fish for sale.   (1)

    iii) Explain your answer.   (2)

6   The diagram shows part of a farmland food chain.

Lettuce   Snail   Thrush   Hawk

If one organism in a food chain is affected in some way, then other organisms in the same food chain will also be affected.

a)   The thrush has many ticks (small parasites) living under its feathers. The ticks are harmful to the thrushes. Write down **three** effects this would have on the food chain.   (3)

b)   The thrushes look for some of their food in gardens and are often killed and eaten by cats. Explain the effect of this on the numbers of snails and lettuces.   (2)

c)   Farmers are encouraged to leave hedgerows around their fields. Suggest **two** reasons why this might increase the population of thrushes.   (2)

7   Deforestation removes many tens of thousands of trees every year.

a)   Nearby farms are often flooded when forests have been cut in this way. Explain why.   (1)

b)   Rainforests are important habitats for many animals. Give **two** reasons why fewer animals can survive if trees have been removed.   (2)

c)   Fallen leaves and fruits from the cut trees can be decomposed (broken down).

    i)   Name the types of organism which carry out this breakdown.   (1)

    ii) During this breakdown, some energy is transferred to the thermal store of energy in the environment. Name the biological process which is responsible for this release of energy during decomposition.   (1)

Some small forest lizards pile up decomposing leaves over their eggs. These incubate the eggs until the young lizards hatch out. The proportion of male and female lizards that hatches is affected by the temperature of incubation. Biologists have collected eggs and incubated them at different temperatures. The results are shown in the following graph.

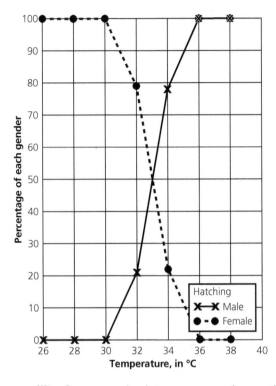

iii) Conservationists want to release the lizards back into a suitable habitat. They would like to release one male for every female. Explain why they need to release males and females in equal numbers. (1)

iv) Use the graph to estimate the temperature at which 50% of the hatching lizards will be male and 50% will be female. (1)

8 Roach are fish that live in freshwater rivers. Local fishers report that they have not caught as many roach as usual from Coney Bridge. They also say that the river smells unpleasant, and there is more algae growing in the water. Scientists from the Environment Agency measure the oxygen levels at different points in the river. Their results are shown below.

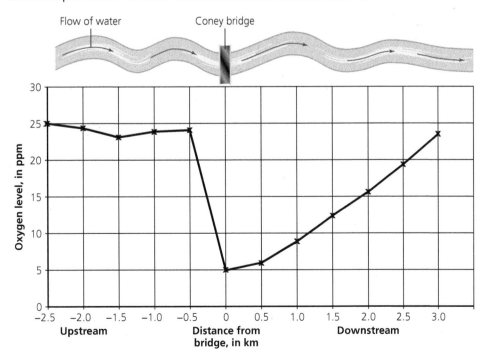

a)   State the oxygen level in the river at Coney Bridge.   (1)

b)   Roach only live in water with oxygen levels higher than 20 ppm. Suggest how far downstream the fishers would be likely to find roach.   (1)

The scientists also collect samples of the river animals found at different points in the river. Their results are shown in this table. (A + means they found the animal at the location; a blank cell means they did not find the animal there.)

Animals collected	Distance from Coney Bridge, in km							
	− 2.0	− 1.5	− 1.0	− 0.5	0	0.5	1.0	1.5
Caddis fly larvae	+	+	+	+				
Bloodworms							+	
Tubifex (Sludge worms)					+	+	+	
Water lice							+	+
Freshwater shrimps	+	+	+	+				
Mayfly nymphs	+	+	+	+				
Stonefly nymphs	+	+	+	+				
Rat-tailed maggots					+	+		

c)   State the name of **one** other animal that only lives in oxygen levels above 20 ppm.   (1)

d)   Name **two** animals that are only found when the oxygen level is below 10 ppm.   (2)

e)   Roach are predators. Suggest **one** reason why the number of roach falls close to Coney Bridge.   (1)

f)   The image shows a rat-tailed maggot.

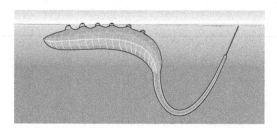

Suggest **one** way that this animal is adapted to living in water with a low oxygen level.   (1)

g)   Suggest how scientists could make a plan to conserve biodiversity in the river.   (3)

9 Variation and classification

1 Choose the option which best completes each of the following.

a) Butterflies are classified as insects because they _____ (1)

lay eggs   **have three main body parts**

can fly   **feed on nectar**

b) The genes that control the characteristics of a cell are part of the

_____ (1)

membrane   **cytoplasm**

nucleus   **chloroplasts**

c) An insect is classified as an arthropod because it _____ (1)

has wings   **has jointed limbs**

has three main body parts   **lays eggs**

d) One of the following human characteristics is not affected by environment. It is

_____ (1)

body mass   **eye colour**

height   **arm strength**

e) Polar bears have thick white fur and live in snowy areas. This is an example of

_____ (1)

adaptation   **variation**

development   **growth**

f) An eagle is classified as a bird because it _____ (1)

has scales   **can fly**

has a beak   **feeds on other birds**

g) Arranging living things into groups of related organisms is called

_____ (1)

fertilisation   **variation**

adaptation   **classification**

h) _____ is an example of discontinuous variation. (1)

The mass of different seeds in a seed pod

Blood group in humans

Height of oak trees

Chest circumference in men

i) Fungi are not included in the plant kingdom because they do not

_____ (1)

reproduce   **respire**

photosynthesise   **excrete**

2   Use the key to identify the coral reef fish shown on the right.

1	Shape is very long and thin	go to 2
	Shape is not long and thin	go to 3
2	Fins are pointed	trumpetfish
	Fins are smooth	eel
3	Eyes on top of head	go to 4
	Eyes on each side of head	go to 5
4	Has a long, thin tail	ray
	Has a blunt tail	plaice
5	Has stripes	go to 6
	Does not have stripes	boxfish
6	Has dark tips to fins and tail	clownfish
	Does not have dark tips to fins and tail	angelfish

3   Sally kept five chickens in her garden.

The table below contains some information about these chickens.

Name of chicken	Sex of chicken	Number of eggs laid per year
Amy	Female	98
Beth	Female	125
Chrissie	Female	105
Dannie	Female	95
Eric	Male	0

a) Sally wants to increase the number of eggs per year by using selective breeding.

   i)   Choose which two chickens she should breed together. (1)

   ii) Sally lets some of the eggs hatch and the chickens become mature. She chooses chickens to breed from among these offspring.

       State which ones she should choose. (1)

b) There are other characteristics of the chickens that might be used during selective breeding.

Choose from this list the **two** characteristics that are the most useful. (2)

- colour of feathers
- amount of milk produced
- size of eggs
- low life expectancy
- size of beak
- resistance to disease

4 Living organisms can be classified according to characteristics they have.

a) Suggest which **three** of the following characteristics are likely to be the most useful for classifying animals. (3)

- the type of skin it has
- how heavy it is
- whether or not it has a bony skeleton
- how fast it can run
- how long it is
- whether or not it has eyes

b) Copy the words in the boxes below and then draw lines to match the descriptions of living organisms with the name of the classification group. (5)

Group	Description of characteristics
Spider	Cells with a definite cell wall but no chlorophyll
Insect	Produces spores and cells contain chlorophyll
Fungus	Two body parts and eight jointed legs
Fern	Body is made of a single cell, with a clear nucleus and cytoplasm
Protist	Three body parts and six jointed legs

5 Samir and Nisar are investigating certain features of other members of their year group at school.

a) The first feature they investigate is whether each pupil has 'joined' ear lobes or 'hanging' ear lobes.

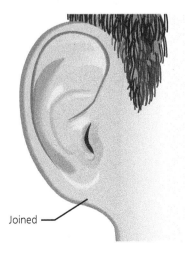

Joined

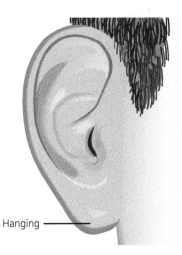

Hanging

They record their results in a table.

Joined ear lobes	Hanging ear lobes
22	8

Use a chart like the one below to draw on a bar to show how many pupils have 'joined' ear lobes.

(2)

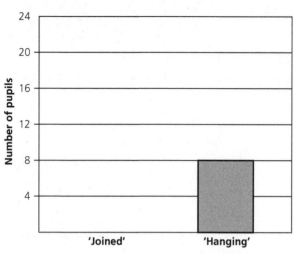

b) Next they investigate the length of the forearm (from the elbow to the tip of the middle finger).

i) Explain why it is important that each pupil keeps their arm straight during the measurement.

(1)

This bar chart shows their results.

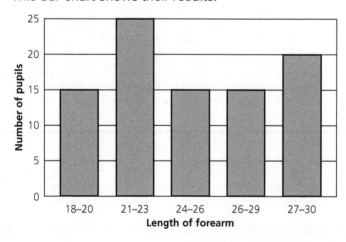

ii) Suggest the units they used for their measurements of forearm length.

(1)

iii) Give **one** mistake in the way they grouped the arm lengths in their bar chart.

(1)

6   The diagram below shows five mammals.

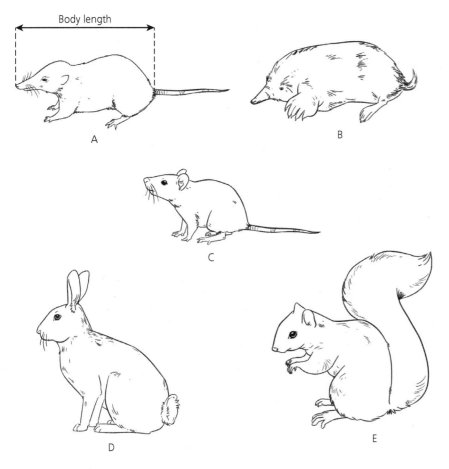

a)   Use the key to identify each of these mammals. Write the letter for each mammal in the
     table that follows.   (5)

1   Tail more than half body length   *go to 2*

    Tail less than half body length   *go to 4*

2   Ears on top of head, with thick tail   *Sciurus carolinensis*

    Ears on side of head, with straight, thin tail   *go to 3*

3   Nose pointed   *Sorex araneus*

    Nose blunt   *Clethrionomys glareolus*

4   Front legs wider than long   *Talpa europaea*

    Front legs longer than wide   *Oryctolagus cuniculus*

Name of mammal	Letter
Clethrionomys glareolus	
Oryctolagus cuniculus	
Sciurus carolinensis	
Sorex araneus	
Talpa europaea	

b) This image shows a young deer feeding from its mother.

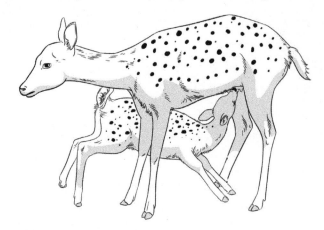

State **two** features of the deer, visible in this image, that distinguish mammals from other vertebrates. **(2)**

7 A scientist working in Scotland believes that otters are very good at catching fish because they can keep their bodies warmer in cold water. She uses very small dataloggers to record the body temperatures of both fish and otters in water at different temperatures. This graph shows her results.

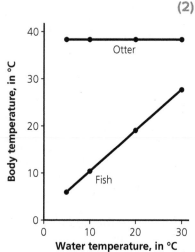

a) Higher body temperatures speed up the action of cells involved in respiration and digestion, and make muscles more flexible. Explain why the scientist thinks that the otters are at an advantage if the water is at 5 °C. **(3)**

b) The water temperature close to the outflow from a power station rises to 25 °C. State what will happen to the body temperature of:

   i) the otter **(1)**

   ii) the fish. **(1)**

c) This image shows an otter.

   i) Visible in the image, give **two** ways in which the otter is adapted for swimming quickly underwater. **(2)**

   ii) The otter is a mammal. Give **one** feature, not shown in the image, that is shared by all mammals. **(1)**

   iii) The otter will eat frogs, grass snakes and even ducklings. Give **one** feature that is shared by otters, fish, frogs, snakes and ducklings. **(1)**

d) The scientist notices that not all of the fish are the same length, even though they look the same otherwise. What is this an example of? **(1)**

10   Particle theory and states of matter

1   Choose the option which best completes each of the following.

   a)   When a solid melts, the particles _____   (1)

     vibrate more   **vibrate less**

     vibrate to the same extent   **stop vibrating completely**

   b)   The change of water from liquid to water vapour is _____   (1)

     precipitation   **freezing**

     evaporation   **condensation**

   c)   During the water cycle, the change of water vapour to droplets of liquid water is

     _____   (1)

     evaporation   **condensation**

     precipitation   **boiling**

   d)   In an experiment to investigate the effect of temperature on evaporation, temperature is

     _____   (1)

     a fixed variable   **the control**

     the dependent variable   **the independent variable**

   e)   Particles move through liquids and gases by _____   (1)

     concentration   **diffusion**

     dilution   **dispersal**

   f)   Density can be defined as _____   (1)

     mass × volume $\dfrac{\textbf{mass}}{\textbf{volume}}$

     $\dfrac{\textbf{volume}}{\textbf{mass}}$ $\dfrac{\textbf{mass}^2}{\textbf{volume}}$

   g)   At the boiling point _____   (1)

     liquid turns into a gas   **a gas turns into a solid**

     a solid turns into a liquid   **a liquid turns into a solid**

   h)   During a change of state, the mass of a substance _____   (1)

     remains the same

     rises

     falls

     may or may not change, depending on the two states involved

2   Sulfur is an element that can exist as a solid, a liquid or a gas. The diagram below shows particles of sulfur in different states, and the letters A, B, C and D represent changes of state between solid, liquid and gas.

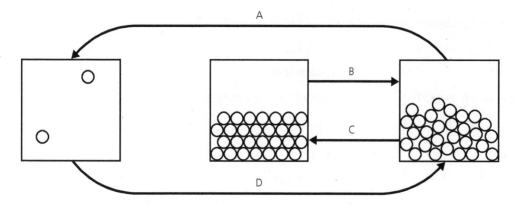

a) Name change of state A.   (1)

b) Name change of state B.   (1)

c) Name change of state C.   (1)

d) Name change of state D.   (1)

3   Choose the option which best completes each of the following.

a) Salty water _____   (1)

   freezes at 0 °C   **freezes above 0 °C**

   doesn't freeze   **freezes below 0 °C**

b) Adding impurities to water causes its boiling point to _____   (1)

   fall   **stay the same**

   rise   **rise to twice its original level**

4   The volume of an object can be found by the displacement of water, and its mass can be found using a weighing machine. Juno wants to find out if her brooch is made of silver and so makes the following measurements.

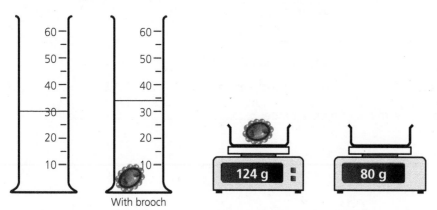

a) Calculate the volume of the brooch. Show your working.   (2)

b) Calculate the mass of the brooch. Show your working.   (1)

c) Calculate the density of the brooch. Show your working.   (2)

d) Silver has a density of 10.2 g/cm³, lead has a density of 11.5 g/cm³ and nickel has a density of 8.9 g/cm³. State whether the brooch is silver or not.   (1)

5   The diagram below shows the water cycle.

a)   Copy and label the diagram to show where evaporation, condensation and rainfall take place.   (3)

b)   The diagram below shows the state of water in different stages of the water cycle and the arrangement of particles.

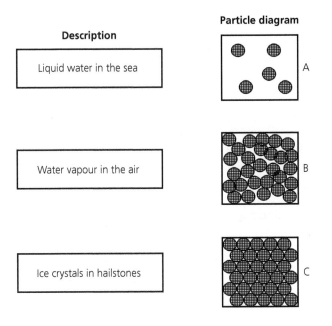

Copy the diagram and draw lines to match the descriptions of water's state with the particle diagrams.   (3)

6 All the pupils in a class decided to celebrate their teacher's retirement. They bought a lot of rubber party balloons and filled half with helium and the other half with air. They then wrapped the balloons in coloured aluminium foil. Each balloon contained exactly the same volume of gas.

a) Explain why the air-filled balloons dropped to the ground but the helium-filled balloons rose. **(2)**

b) The diagram below shows a number of arrangements of particles.

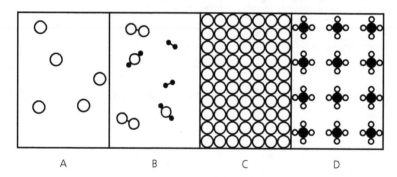

State which letter represents:

i) the helium gas **(1)**

ii) the aluminium foil. **(1)**

c) Over a period of a week, the balloons shrank because the particles of gas escaped. The helium-filled balloons shrank more quickly.

i) Name the process by which the particles of gas moved out of the balloons. **(1)**

ii) Suggest why helium escapes from a balloon more quickly than air does. **(1)**

7 The image shows a can of pressurised propane, used as a fuel for burning paint from wooden window frames. Most of the propane in the can is in the liquid state, but some of it is gas.

a) **i)** Describe what happens to the pressure inside the can when the temperature falls. Explain your answer. **(2)**

ii) Describe what would happen to the flame from the paint burner if the painter were to use it outdoors on a cold day. **(1)**

iii) There is a warning on the can not to throw it into a fire. Explain why this is important. **(1)**

b) State which of these statements about the propane liquid in the can are correct. (3)

The molecules of the liquid are:
- smaller than those in the gas
- the same distance apart as those in the gas
- closer together than those in the gas
- moving faster than those in the gas
- the same size as those in the gas
- bigger than those in the gas
- further apart than those in the gas
- moving more slowly than those in the gas

8 Some solid wax was slowly warmed in a boiling tube. The temperature of the wax during the warming process was measured using a temperature sensor connected to a datalogger.

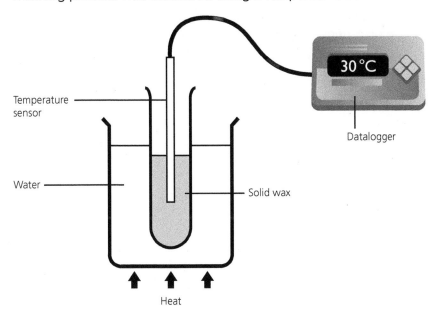

Here is a graph of the results.

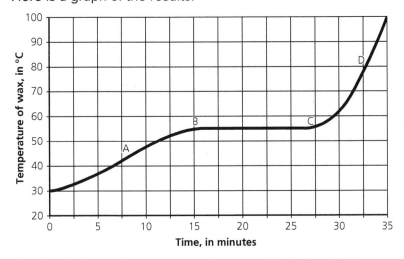

a) i) State at what point on the graph (A, B, C or D) the wax began to change state. (1)

ii) Explain your answer. (2)

iii) Name the physical state of the wax at point D. (1)

b) Suggest **two advantages** of carrying out the investigation using a temperature sensor and datalogger rather than a thermometer. (2)

c) This particular wax boils at 275 °C. Explain why the wax cannot boil during this investigation. (1)

d) Describe and explain the differences between the movement of the water molecules at the start and the end of the investigation. (2)

9   Robert Brown observed pollen grains moving around on the surface of cold water. This is called Brownian motion.

A teacher set up a model to show Brownian motion. It is shown in the diagram below. The ping-pong balls will not roll off the speaker because it has a raised edge.

a)   The teacher turned on the power supply. The loudspeaker started to vibrate. Predict what will happen to the ping-pong balls and small balloon.   **(1)**

b)   State what the ping-pong balls and small balloon represent in this model of Robert Brown's observations.   **(2)**

c)   Use the model to explain why Brown observed that the pollen grains moved.   **(3)**

d)   Suggest how the process of Brownian motion can be used to explain diffusion.   **(2)**

11   Atoms, elements and compounds

1   Choose the option which best completes each of the following.

    a)   A typical metal _____   (1)

        is not a good conductor of electricity

        does not have a high melting point

        is shiny in appearance

        is not malleable

    b)   A substance made of only one type of atom is _____   (1)

        a gas   **an element**

        likely to be of low density   **an alloy**

    c)   The particles in a liquid are likely to be _____   (1)

        in a regular fixed pattern   **not moving**

        moving slightly   **far apart**

    d)   Copper reacts with oxygen to produce _____   (1)

        a gas   **an oxide**

        an acid   **an element**

    e)   An element that can conduct electricity is _____   (1)

        oxygen   **iron**

        sulfur   **nitrogen**

    f)   The simplest particles found in matter are _____   (1)

        molecules   **compounds**

        atoms   **elements**

    g)   The particles in a compound are called _____   (1)

        atoms   **elements**

        molecules   **reactants**

    h)   The chemical symbol for iron is _____   (1)

        I   **Fe**

        Ir   **In**

    i)   The number of elements in sulfuric acid, H_2SO_4 is _____   (1)

        1   **6**

        3   **8**

    j)   The approximate proportion of oxygen in air is _____   (1)

        0.03%   **21%**

        6%   **79%**

2   The images show different elements that have been used to produce different objects.

a)   Copy the words in the boxes below and then draw lines to match the element to the reason for using it.   (5)

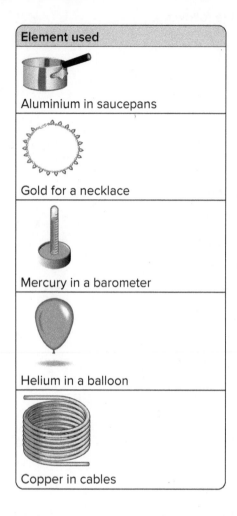

Element used
Aluminium in saucepans
Gold for a necklace
Mercury in a barometer
Helium in a balloon
Copper in cables

Reason for choosing this element
It is lighter than air
It is a good conductor of heat
It conducts electricity and is easy to stretch
It stays shiny because it does not react with oxygen in air
It stays liquid at room temperature

b)   Elements have a number of different properties.

i)   Name **two** properties of copper that would make it suitable for use in the solid base of expensive saucepans.   (2)

ii)   Name a property of helium that might make it suitable for storage in metal cylinders.   (1)

3   Scientists were collecting material from the bottom of the Pacific Ocean. One of them believed that they had found a new metallic element – they called it oceanium.

a)   State which **two** properties from the list below suggest that oceanium could be a metal.   (2)

   ● It glows in the dark.
   ● It is a green solid.
   ● It has a high melting point.
   ● It is a good conductor of heat and electricity.
   ● It does not stick to a magnet.

b)   One of the scientists tried adding some dilute acid to the solid. The result was a blue solution with green solids floating in it. Name **one** method he could use to separate the floating green solids from the solution.   (1)

4   This list shows a number of properties of different materials.

A   non-conductor of electricity

B   poor conductor of heat

C   magnetic

D   can be compressed

E   good conductor of heat

F   very flexible

G   very high melting point

H   good conductor of electricity

Match the properties with the statements below by selecting letters to fill the blanks.   (5)

a)   _____ makes plastic a good material for the handle of a kettle.

b)   _____ makes it possible to pump a lot of air into a bicycle tyre.

c)   _____ makes cotton a good material for shoelaces.

d)   _____ and _____ are two properties that make aluminium a good material for cooking pans.

5 A group of pupils carried out a class practical in which they burned magnesium in air.
   They wished to record the mass of magnesium used and the mass of the product formed.

   a) Draw and label a set of apparatus they might use to carry out this investigation. (3)

   b) The results are shown in the table below.

Pupil	Mass of magnesium, in g	Mass of product, in g
Jamal	6.4	10.4
Sam	3.8	6.5
Neela	4.8	8.4
Anja	6.1	10.7
James	2.7	4.0
Billy	4.2	7.0

   i) State the name of the product that is formed. (1)

   ii) The results from the table above are presented on a grid below.

       Use the graph to predict the mass of magnesium required to provide 7.5 g of product. (1)

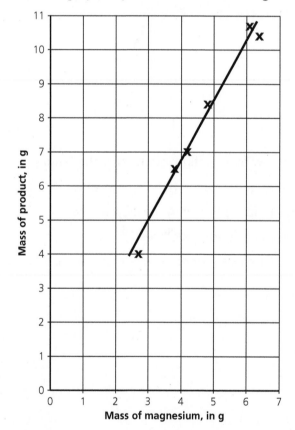

   iii) Use the graph to predict how much product will be formed if 5.5 g of magnesium is
        burned in air. (1)

   iv) Give **one** conclusion you can draw about the relationship between the mass of
        magnesium burned and the mass of product formed. (2)

6   John Dalton used symbols like the ones shown here to represent atoms.

Some possible combinations of these atoms are shown in these diagrams.

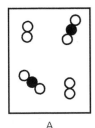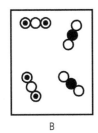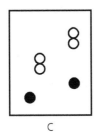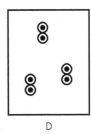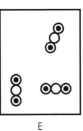

A   B   C   D   E

a) i)   Give the letter that shows a mixture of an element and a compound.   (1)

    ii) Give the letter that shows only a compound.   (1)

b) Give **one** difference between a compound and a mixture.   (1)

c) i)   Suggest a name and formula for the substance represented in D.   (1)

    ii) Suggest names and formulae for the substances represented in C.   (2)

7   The table shows the melting points and boiling points of some elements.

Element	Melting point, in °C	Boiling point, in °C
Sodium	98	883
Mercury	−39	357
Iron	1538	2861
Oxygen	−219	−183
Gold	1064	2856

a) State which element is liquid at room temperature (approximately 21°C).   (1)

b) State which element is a non-metal.   (1)

c) Magnesium chloride is formed when magnesium and chlorine combine together in a chemical reaction. Write the symbols for magnesium and chlorine.   (2)

d) The formula for a substance is FeS. State the name of this substance.   (1)

e) Calcium burns brightly in oxygen, forming calcium oxide (CaO). Calcium oxide reacts with water, forming a compound with the formula $Ca(OH)_2$.

   i)   Name the compound with the formula $Ca(OH)_2$.   (1)

   ii) This compound is slightly soluble in water. Predict the colour of universal indicator when mixed with the solution.   (1)

f) State how many atoms of iron and oxygen there are in the formulae of these compounds: FeO and Fe_2O_3.   (2)

Compound	Number of atoms of iron	Number of atoms of oxygen
FeO		
Fe_2O_3		

1 Choose the option which best completes each of the following.

a) Pure water is _____ (1)

    a solution   **a mixture**

    a compound   **an element**

b) A mixture made of a solvent and an insoluble substance can be

    _____ (1)

    a solution   **an oil**

    a suspension   **a solute**

c) The change of state from liquid to gas is _____ (1)

    condensation   **distillation**

    melting   **evaporation**

d) A separation method that separates a solid from a liquid by careful pouring is

    _____ (1)

    distillation   **decanting**

    filtration   **evaporation**

e) An example of a mixture is _____ (1)

    iron filings   **water**

    air   **sodium chloride**

f) A pure substance _____ (1)

    contains particles of only one type

    contains only atoms

    contains different elements arranged in different ways

    cannot form part of a mixture

g) Carbon dioxide is not _____ (1)

    a gas that makes up about 0.04% of the atmosphere

    a product of photosynthesis

    produced when energy is released in aerobic respiration

    a compound

h) Oxygen gas is not _____ (1)

needed for combustion to take place

needed for aerobic respiration

produced during photosynthesis

a compound

i) A condenser is a piece of apparatus used for separation of _____ (1)

a solvent from a solution	**several different soluble substances**
a solid from a liquid	**gases from the air**

j) The change of state from vapour to liquid is _____ (1)

condensation	**distillation**
melting	**evaporation**

2 The following are different ways of separating the components of mixtures:

A simple distillation

B with a magnet

C chromatography

D filtration

Choose **one** of the letters to show which is the best method to obtain:

a) iron from a mixture of iron filings and sulfur (1)

b) water from seawater (1)

c) chalk from a mixture of chalk and water (1)

d) food colourings from sweets. (1)

3 Pedro had a leaking ballpoint pen, which left a stain on his trouser pocket.
He rubbed some ethanol on to the stain with a tissue, and noticed that the tissue developed a purple stain and the stain on his trousers became lighter.

a) Explain how the ethanol helped to remove the stain. (2)

b) The teacher asked Pedro whether he thought that the ballpoint ink was made of one type of dye or several.

i) Name the technique that Pedro could use to find out. (1)

ii) Pedro obtained the following results:

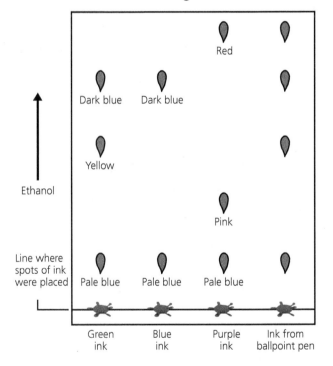

State which colours were present in the ballpoint ink. (1)

iii) State how many coloured substances there were in the blue ink. Explain your answer. (1)

iv) Choose which word is the best one to describe the coloured substances. (1)

● solution
● solvent
● solute
● suspension

4   Sara and David are investigating the effect of solute
    concentration on the boiling point of water. They measure
    out different masses of salt and dissolve each sample in a
    different 500 cm³ of water. They then measure and record the
    temperature at which the water boils.

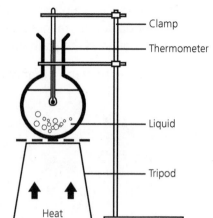

a) i)   Name the **independent variable** in their investigation. (1)

    ii) Name the **dependent variable** in their investigation. (1)

    iii) Sarah and David have a list of possible **fixed** (controlled)
         variables. From this list, choose **two** that must be
         controlled, and **one** that would have little or no effect
         on their results. (3)
         ● volume of water
         ● type of salt dissolved in water
         ● starting temperature of water
         ● room temperature

b) They write down their results in this table.

Mass of salt added, in g	Boiling point, in °C
0	100.0
10	100.6
20	103.0
30	104.0
40	106.0
50	result lost
60	109.2

They then plot their results on a grid.

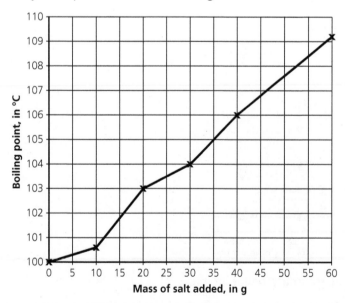

i)   From their graph, suggest a likely result for the one, at 50 g of salt added, which they lost. (1)

ii) Predict the temperature at which a solution of 100 g of salt in 500 cm³ of water
     will boil. Show your working. (2)

5   Su-Yin adds some copper sulfate crystals to a beaker of water.

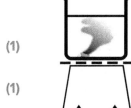

a) i) Describe how she can see that some of the crystals have
        dissolved in the water.   (1)

    ii) Suggest what she can do to increase the amount of the
        copper sulfate that dissolves.   (1)

b) Suggest how Su-Yin can collect copper sulfate crystals from the
    solution in the beaker.   (1)

c) Su-Yin carries out an experiment to investigate how much of three food flavourings will
    dissolve in water at different temperatures. The results are shown in the graph below.

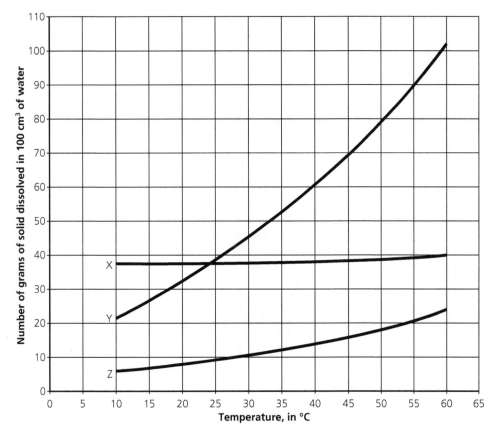

    i)   State how many grams of flavouring Z dissolved in the water at 40 °C.   (1)

    ii) Name the flavouring which dissolved best at 20 °C.   (1)

    iii) Name the **two** flavourings which are equally soluble at 24 °C.   (1)

6   Abi adds some sugar to 100 cm³ of cold water in a beaker. She stirs the water to dissolve the sugar, and then adds more sugar until no more will dissolve.

She repeats the experiment with salt, curry powder and instant coffee. Each time, she uses a different beaker containing 100 cm³ of cold water. The results are shown in the bar chart below.

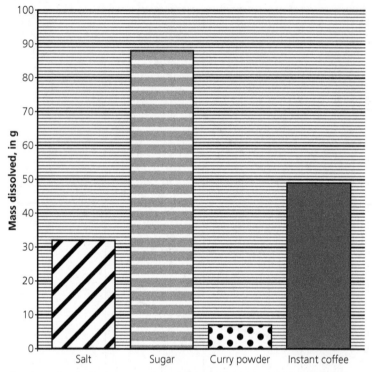

a) i)   State **two** ways in which Abi made this a fair test.   (2)

    ii) State which substance was the least soluble in water.   (1)

    iii) Calculate how many times more soluble sugar was than salt. Give your answer to
         1 decimal place and show your working.   (3)

b) When Abi makes a cup of coffee for her father, the coffee dissolves much more quickly.
    Explain why.   (1)

7   A science teacher sets up the apparatus shown, with a solution of a water-soluble food dye in the flask.

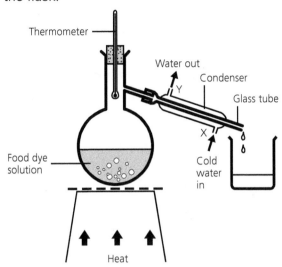

a) i)   State the name of the separation process which the teacher is demonstrating.   (1)

    ii) State which pair of processes will occur during the demonstration. Choose from this list: (1)

        ● melting then evaporation
        ● condensation then evaporation
        ● melting then boiling
        ● evaporation then condensation

b) The teacher says that the liquid that is collected in the beaker is pure water. Describe **one** test that would show the liquid is water and another test that would show that it is pure water.   (2)

c) i)   Water at 20 °C enters the condenser at X. Predict the temperature of the water when it leaves the condenser at Y.   (1)

    ii) Explain your answer.   (1)

    iii) Give **two** ways in which the water vapour is changed as it passes down the glass tube in the condenser.   (2)

8   Copy the words in the boxes below and then draw lines to match the terms with their correct definitions.   (3)

Term	Definition
Concentrated	A mixture of a solvent and a solute
Solution	The liquid part of a solution
Solvent	A solution with many solute particles in a small volume of solvent
Solubility	The amount of a substance that will dissolve in a liquid
Soluble	Able to dissolve

13   Chemical reactions

1   Choose the option which best completes each of the following.

  a)   Each of the following is a sign that a chemical reaction has taken place except for

     _____   (1)

temperature rises	**a new substance being formed**
the reaction being reversible	**fizzing often occurring as a gas is formed**

  b)   The products of fermentation do not include _____   (1)

carbon dioxide	**glucose**
heat	**ethanol**

  c)   If a hydrocarbon is burned in a good supply of air, the products are

     _____   (1)

carbon dioxide + water	**carbon + water**
carbon dioxide + carbon monoxide	**carbon monoxide + water**

  d)   Hydrochloric acid and magnesium oxide react together to produce magnesium chloride

     and water. This type of reaction is an example of _____   (1)

reduction	**neutralisation**
oxidation	**recombination**

  e)   Each of the following is a fossil fuel except for _____   (1)

natural gas	**coal**
oil	**wood**

  f)   Spoilage of food is an example of a harmful chemical reaction. Food cannot be preserved

     by _____   (1)

adding extra water	**drying**
removing oxygen	**keeping food in acidic conditions**

  g)   _____ is not a chemical reaction.   (1)

Photosynthesis	**Separation of a mixture of iron filings and sulfur**
Respiration	**Burning natural gas in air**

2   Copy the boxes below and then draw lines to match the chemical reactions with their
possible uses.   (4)

Reaction	Use for the reaction
iron oxide + carbon monoxide → carbon dioxide + iron	For plants to make food
methane + oxygen → carbon dioxide + water	To extract a metal from its ore
water + carbon dioxide → glucose + oxygen	To release energy in living organisms
glucose + oxygen → carbon dioxide + water	To transform chemical energy to thermal energy for heating

3   A science teacher heated iron filings and sulfur on a tray.

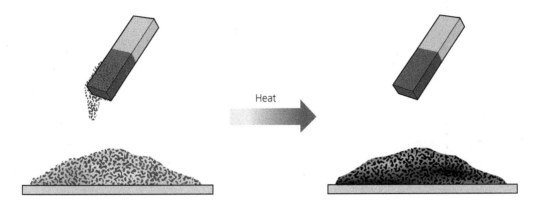

Heat

a) i)   Write out a word equation for this reaction.   (1)

   ii)   From the information in the diagram, give **one** piece of evidence that a chemical
         reaction has occurred.   (1)

b) i)   Suggest the name and the formula for the solid formed when zinc is heated with sulfur.   (2)

   ii)   Some fossil fuels contain sulfur. When fossil fuels burn, sulfur reacts with oxygen.
         Write out a word equation for this reaction.   (1)

   iii) State what type of chemical reaction this is.   (1)

4   This apparatus can be used to burn magnesium ribbon in
    air. The process is demonstrated by a science teacher.

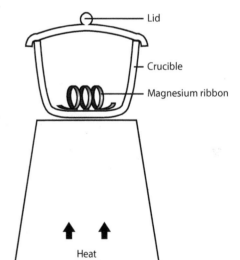

a) i)   Explain why the teacher tells the pupils
         that they should never look directly at
         burning magnesium.   (1)

   ii)   The teacher wants to make a note of the
         changes in mass during the reaction. They
         rub the magnesium ribbon with a
         piece of abrasive paper before beginning
         the experiment. Explain why this is
         important.   (1)

b) The following results are obtained:

Mass of crucible, in g	50	50	50	
Mass of crucible + magnesium ribbon, in g	63	61	62	
Mass of crucible + contents after heating, in g	69.8	70.1	70.1	

   i)   Copy and complete the table by writing the mean values for the measurements in the
        empty column.   (1)

   ii) Explain these results.   (2)

   iii) Write out a word equation for the reaction that has taken place during the
        demonstration.   (1)

5   Choose the option which best completes each of the following.

   a)   A substance that will not decompose on heating is _____   (1)

   copper sulfate   **calcium carbonate**

   copper   **copper carbonate**

   b)   The equation that represents fermentation (anaerobic respiration in yeast) is

   _____   (1)

   sugar → lactic acid + energy

   sugar → alcohol + carbon dioxide + energy

   alcohol + carbon dioxide → sugar + energy

   sugar → alcohol + energy

6   The table below contains some information about the sources and effects of greenhouse gases.

Name of gas	Sources of gas	Percentage overall contribution to the greenhouse effect
Methane		14
CFCs	Aerosols, refrigerants and coolants	21
	Burning forests and fossil fuels, manufacture of cement	
Nitrogen oxides	Breakdown of fertilisers, burning fuel in internal combustion engines	7
Low-level ozone	Combination of nitrogen oxides with oxygen	2

   a)   Copy and complete the table, by writing in the empty boxes:

   • the source of methane
   • the main greenhouse gas
   • the percentage contribution this gas makes to the greenhouse effect.   (2)

   b)   Suggest **three** possible harmful results of the greenhouse effect.   (3)

   c)   Burning fossil fuels also causes air pollution, including acid rain. Lakes that have been acidified by acid rain have very little remaining aquatic life. Some lakes have been treated by adding large quantities of calcium hydroxide, which quickly dissolves in the lake water.

   i)   State the effect this will have on the pH of the lake.   (1)

   ii)   When the calcium hydroxide reacts with the sulfuric acid in the lake, a salt is formed. Name this salt.   (1)

   iii) Write a word equation for the reaction which produces this salt.   (1)

7 A manufacturer of model figures was interested in changing the material used to cast the models.
The manufacturer mixed two components of the resin, and then left the resin to swell. The resin needs to swell before it can be poured – it only hardens when it is baked above 200°C. This diagram shows the changes in volume of the resin over a 30-minute period.

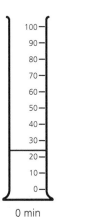 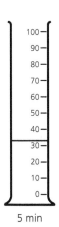 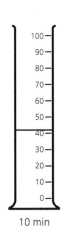 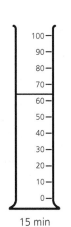 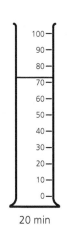 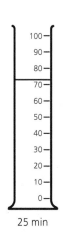 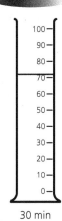

| 0 min | 5 min | 10 min | 15 min | 20 min | 25 min | 30 min |

(Start of experiment)

a) i) Copy and complete the table below, using the manufacturer's results. (2)

Time, in minutes	Volume of resin mix, in cm³
0	
5	
10	
15	
20	
25	
30	

ii) Draw a line graph of these results on a graph grid like the one below. (3)

iii) The manufacturer believed that he would get the most stable resin if he only allowed the mix to swell to two and a half times its original volume. Use the graph to predict how long he would need to leave the mix to achieve this result. (1)

b) He decided to add different quantities of a hardener to see the effect on the volume of the resin mix.

i) Name the **independent variable** in this investigation. (1)

ii) Name the **dependent variable** in this investigation. (1)

iii) Suggest **two** other variables which the manufacturer would need to control if this were to be a fair test. (2)

c) One feature of the hardened resin is that it does not react with air or with water. Suggest why this is important in the finished models. (1)

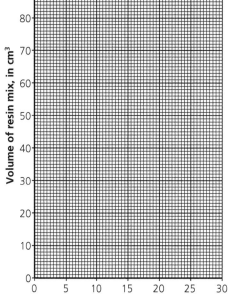

8   Two factors affect the heat delivered by a Bunsen burner – the supply of gas from the gas tap and the supply of air through the air hole.

    a)   The hottest Bunsen flame is achieved when _____   (1)

        the gas tap is fully on and the air hole closed

        the gas tap is fully on and the air hole half open

        the gas tap is half on and the air hole half open

        the gas tap is fully on and the air hole fully open

    b)   The hottest part of the Bunsen burner flame is _____   (1)

        the cone of the flame   **the blue part of the flame**

        inside the barrel   **just behind the open air hole**

9   A student placed an iron nail inside a $25\,cm^3$ measuring cylinder. She added boiled water until the meniscus reached the $15\,cm^3$ mark, then sealed the measuring cylinder. Within a few days the nail began to show signs of rust, and the water level began to rise.

    a)   Explain why the water level began to rise.   (1)

    b)   Explain why boiled water was used in this experiment.   (1)

    After a long period, the nail was very rusty and the water level stopped rising.

    c)   Predict the level to which the water had now risen.   (1)

    d)   Explain your prediction.   (1)

    The student carefully removed the nail from the water and dried it.

    e)   The nail would be _____   (1)

        heavier than at the start of the experiment

        lighter than at the start of the experiment

        the same weight as at the start of the experiment

    f)   Explain your answer to **e)**.   (1)

10 The following paragraph is about damage to stone buildings. Copy and complete it by filling in the missing words, chosen from this list.   (5)

acid rain   sulfur dioxide   combustion   granite   nitrogen   sulfuric acid   limestone

_____ of fossil fuels produces the gas _____. This gas dissolves in water in the atmosphere to produce _____, which falls as _____. This compound can cause damage to buildings, especially those made from _____.

14 The reactions of metals

1 Choose the option which best completes each of the following.

a) A metal that is unreactive and so would be suitable for electrical contacts is

_____ (1)

iron   magnesium

aluminium   silver

b) When copper carbonate is heated, the reaction that takes place is

_____ (1)

neutralisation   displacement

decomposition   combustion

c) When copper is heated strongly in an open dish its mass will

_____ (1)

increase   decrease

stay the same   increase then decrease

d) The chemical symbol for sodium is _____ (1)

So   Na

S   Su

e) Metals are not _____ (1)

often shiny   good conductors of heat

malleable   usually of low density

f) The following elements all conduct electricity, but _____ is not
a metal. (1)

iron   copper

graphite   nickel

g) Metal railings can rust, so they must be made of _____ (1)

aluminium   lead

copper   iron

h) _____ must be present for metal railings to rust. (1)

Carbon dioxide and oxygen   Water and carbon dioxide

Oxygen and water   Water and hydrogen

2   Remi is investigating the reaction between metals and hydrochloric acid. He adds 20 cm³ of hydrochloric acid to each of five test tubes, then places equal-sized pieces of metal into four of the tubes.

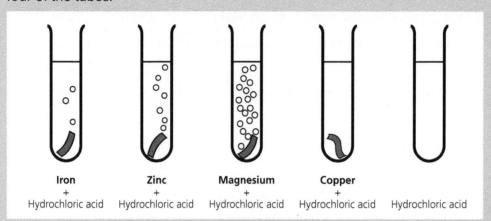

Iron + Hydrochloric acid   Zinc + Hydrochloric acid   Magnesium + Hydrochloric acid   Copper + Hydrochloric acid   Hydrochloric acid

a) i)   Identify the **independent variable** in this investigation.   (1)

    ii) Identify the **dependent variable** in this investigation.   (1)

    iii) Suggest how the **dependent variable** is measured.   (1)

    iv) Name **two** steps that Remi takes to ensure that this is a fair test.   (2)

b) Remi wants to know where gold would fit into this series, but his teacher says that gold is too expensive.

    i)   Predict the result that Remi would have obtained if he had been allowed to use gold in his investigation.   (1)

    ii) Explain your answer.   (1)

3   Ahmed is concerned that his bike is rusting, so he tries to work out the conditions for rusting. He sets up five test tubes containing iron nails, as shown below.

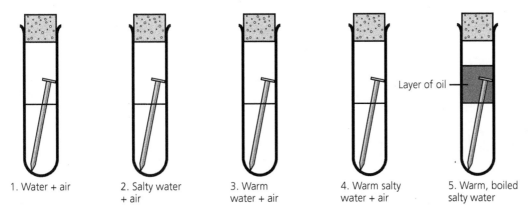

1. Water + air   2. Salty water + air   3. Warm water + air   4. Warm salty water + air   5. Warm, boiled salty water

He presents his results as a bar chart, shown below.

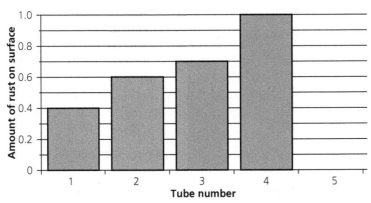

a) i)   Copy the bar chart and insert the missing bar for tube 5.   (1)

         Give a reason for the size of the bar.   (1)

    ii) State which has the bigger effect on rusting – salt or warmth.   (1)

         Give a reason for your answer.   (1)

b) Ahmed sets up a sixth test tube: this one contains a nail in vinegar. He notices that the iron nail reacted with the vinegar.

    i)   State whether vinegar is acidic, alkaline or neutral.   (1)

    ii) During the reaction, bubbles of gas are given off. Name this gas.   (1)

    iii) If the gas could be collected, suggest how you could test your answer.   (1)

c) Zinc is sometimes used to galvanise metal surfaces and prevent rusting.

    i)   Explain how zinc prevents rusting.   (2)

    ii) Suggest why galvanising is not used on bicycle chains.   (1)

4   In the extraction of iron from iron ore, haematite (iron oxide) is reacted at high temperature with coke (carbon) in the presence of oxygen. The oxygen combines with the carbon to form carbon monoxide.

   a)   Copy and complete this word equation for the reaction between carbon monoxide and iron oxide.   (3)

   _____ + _____ → _____ + carbon dioxide

   b)   The cast iron produced in this way is brittle, and is usually modified in some way to make it more useful. The table below shows the percentage of carbon in four different materials.

Material	Percentage of carbon
Cast iron	4.0
Wrought iron	0.2
High-carbon steel	0.8
Mild steel	0.4

   i)   High-carbon steel is used for some knife blades. Calculate what proportion of the carbon must be removed from cast iron to produce high-carbon steel. Show your working.   (2)

   ii)   The highest-quality knife blades are made from stainless steel. This is made by removing most of the carbon and adding small amounts of other metals. State the main **advantage** of stainless steel.   (1)

   c)   The graph below shows how the percentage of carbon affects the strength of the materials in the table: the strength is measured using a forcemeter.

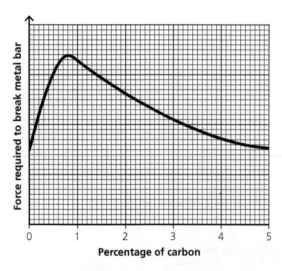

   i)   State the percentage of carbon in the material with the greatest strength.   (1)

   ii)   State which is the strongest material in the table in **b)**.   (1)

5 a) Copper is best described as _____

    an element   **a compound**

    a metal   **a metallic element**   (1)

  b) Metals are usually found as compounds and must be extracted from these compounds before they can be used.

    i) State the name given to these naturally occurring metal compounds.   (1)

    ii) Choose which **one** of these metals is most usually found in its 'free' state (i.e. not in a compound).   (1)

        iron   **copper**

        silver   **aluminium**

  c) An iron nail is placed into some silver nitrate solution. A reaction takes place between the iron and the silver nitrate. Copy and complete this word equation for the reaction.   (3)

    iron + _____ ⟶ _____ + _____

  d) Gold does not tarnish (go dull), but aluminium often does. Explain why.   (1)

1 Choose the option which best completes each of the following.

a) A substance that can attack other materials, including human skin, is

_____ (1)

| hot | corrosive |
| strong | an indicator |

b) A chemical reaction between an alkali or base and an acid is

_____ (1)

| neutralisation | oxidation |
| indication | decomposition |

c) A solution with a pH of 7 is _____ (1)

| acidic | basic |
| corrosive | neutral |

d) One product of a neutralisation reaction is _____ (1)

| an acid | an alkali |
| a salt | an indicator |

e) A compound that reacts with an acid to release carbon dioxide is

_____ (1)

| a base | a carbonate |
| carbon monoxide | a metal chloride |

f) A gas that turns limewater milky is _____ (1)

| hydrogen | nitrogen |
| oxygen | carbon dioxide |

g) Acids react with metals to produce _____ (1)

| salt + water | salt + hydrogen |
| water vapour | salt + carbon dioxide |

h) An alkali will always turn _____ (1)

| litmus paper red | universal indicator yellow |
| litmus paper blue | limewater milky |

i) The chemical formula NaOH represents _____ (1)

| nitrogen hydroxide | sodium hydroxide |
| sodium chloride | sodium hydride |

2   The pH of a soil sample can be tested by shaking the soil with water, letting the particles settle and then adding universal indicator solution.

This table shows the pH values of six soil samples.

Sample	pH value
A	8.0
B	7.5
C	7.0
D	8.0
E	4.5
F	6.0

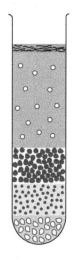

a)   State which soil sample would give a green colour with the universal indicator solution.   (1)

b)   Cauliflower and sprouts grow better in alkaline soil. State which soil samples should allow cauliflower and sprouts to grow well.   (1)

c)   Rhododendron is a shrub which grows better in acidic soils. State which of the soil samples would allow the rhododendron to grow well.   (1)

d)   Crushed and heated limestone can be dissolved in water to produce calcium hydroxide. This calcium hydroxide (slaked lime) is sometimes added to acidic soils. Name the type of reaction that takes place between the lime and the soil.   (1)

3   Hydrochloric acid is an example of a strong acid, and is present in the stomach during the digestion of food.

a)   A student adds five drops of hydrochloric acid to a small volume of universal indicator solution in a test tube.

   i)   State which colour the mixture of indicator and hydrochloric acid is likely to be.   (1)

   ii)   Suggest a likely value for the pH of the hydrochloric acid.   (1)

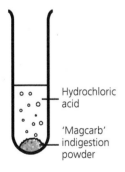

Hydrochloric acid

'Magcarb' indigestion powder

b)   Too much hydrochloric acid produced in the stomach can be a cause of indigestion. 'Magcarb' indigestion tablets contain magnesium carbonate, and can be crushed into a powder. If the powder is added to hydrochloric acid in a test tube, the mixture will fizz.

   i)   Copy and complete the word equation for the reaction which occurs.   (2)

   hydrochloric acid + _____ ➔ _____ + _____ + water

   ii)   Use this equation to explain why the mixture fizzed after addition of the powdered tablets.   (1)

   iii) The student continues to add powder to the acid, and notices that the mixture stops fizzing. Explain why the fizzing stops.   (1)

c) Choose which of the following words could be used to describe magnesium carbonate. (2)

a solvent   **an element**   **a salt**

an indicator   **a mixture**   **a compound**

4   Wasps and bees are both insects that defend themselves using stings. The 'sting' involves the injection of a solution through the skin. Bee stings are pH2 and wasp stings are pH10.

a) Copy and complete the table below to show whether the stings are acid or alkaline, and suggest what colour they would change universal indicator solution to. (4)

Type of sting	Acid or alkaline	Colour of indicator solution
Wasp sting		
Bee sting		

b) Some common household substances can be used to neutralise wasp and bee stings. Six of these substances are shown in the table below.

Substance	pH value
Water	7
Washing soda	11
Baking soda	8
Bicarbonate toothpaste	8
Vinegar	5
Lemon juice	3

Give the name of **one** substance in the table that could neutralise:

i)   a wasp sting (1)

ii) a bee sting. (1)

c) Explain why it is useful that toothpaste is slightly alkaline. (2)

d) Nettle leaves contain small cells that release formic acid.

i)   Suggest why some leaves can be used to get rid of the irritation from a nettle sting. (1)

ii) Explain why the relief from the sting is quicker if the leaf is crushed up before it is used in this way. (1)

5   Scientists believe that acid rain is caused by gases in the atmosphere. An investigation into the formation of acid rain was carried out by collecting a number of gases. Each of the gases was bubbled through a sample of green, neutral universal indicator solution.

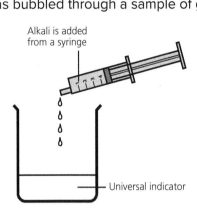

Alkali is added from a syringe

Universal indicator

a)   Three of the gases caused the indicator to change colour. Once all the gas samples had been bubbled through the indicator, alkali was added, from a syringe, until the universal indicator changed back to green.

The results are shown in the table below.

Gas collected	Change in colour of indicator	Volume of alkali needed to return indicator to green colour, in cm³
Air	No change	0
Carbon dioxide from burning coal	Green to red	6.9
Exhaust gases from an idling car	Green to red	1.2
Methane from a refuse tip	No change	0
Human breath	Green to yellow	0.2

   i)   State which gases formed neutral solutions.   (2)

        Explain your answer.   (1)

   ii) State which gas produced the most acidic solution.   (1)

        Explain your answer.   (1)

   iii) State the name given to a reaction between an acid and an alkali.   (1)

b)   i)   Some metals used in buildings may react with substances in the air. One possible
          reaction is shown below. Copy and complete the word equation.   (2)

          copper + carbonic acid ➔ _____ + _____

     ii) Bronze statues often change to a green colour after many years of exposure to the
          air. Use this word equation to explain why.   (2)

6   Jack placed a conical flask on a pan balance. The flask weighed 100 g. He added 50 g of dilute hydrochloric acid and 5.0 g of calcium carbonate to the flask. The total mass of the flask and its contents was 155 g.

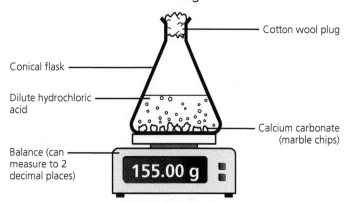

Cotton wool plug

Conical flask

Dilute hydrochloric acid

Calcium carbonate (marble chips)

Balance (can measure to 2 decimal places)

155.00 g

a)   The calcium carbonate and the hydrochloric acid quickly reacted together. Copy and complete the word equation for the reaction which took place.   (3)

calcium carbonate + hydrochloric acid   → _____ + _____ + _____

b)   When the reaction stopped, the total mass had decreased from 155.00 g to 152.70 g. Some water had evaporated from the flask. Suggest another reason for the mass to fall in this way.   (1)

c)   Explain how Jack knew that the reaction had stopped.   (1)

d)   At the end of the reaction, the calcium carbonate had neutralised the acid. Jack tested a few drops from the flask with universal indicator paper. Suggest the colour of the universal indicator paper after this test.   (1)

e)   Calcium carbonate is a very common compound. State which of the materials in this list are mainly calcium carbonate.   (3)

marble   **chalk**   **limestone**   **coal**   **sand**   **glass**

f)   Metals also react with acids.

i)   Name the gas which is produced when a metal reacts with an acid.   (1)

ii)   Describe a test for this gas.   (1)

7   Sodium hydrogencarbonate (often called bicarbonate of soda) is present in baking powder.

a)   State which of these properties of sodium hydrogencarbonate are especially important for its use as baking powder.   (2)

it is very soluble in water   **it can be kept in cardboard boxes**

it is not poisonous   **it is a white solid**

b)   Baking powder also contains citric acid. When the baking powder is added to water, the acid and the hydrogencarbonate react together. Explain how this helps to give a light texture to cakes.   (1)

16 Energy resources and transfers

1 Energy can be stored in a number of different types of energy store.

This table lists energy stores and their definitions, but the two columns have become mixed up. Match the letter for each energy store with the number for its definition.

Store of energy	Definition
A Chemical store of energy	1 A moving object will have this store of energy
B Gravitational potential store of energy	2 Energy store in an object due to the internal movement and position of its particles
C Kinetic store of energy	3 Energy store in an object when it is squashed, squeezed or stretched
D Internal (heat/thermal) store of energy	4 Energy store in chemical bonds
E Elastic store of energy	5 Energy store in the attractions between the particles in an atomic nucleus
F Nuclear store of energy	6 Energy store in an object due to its position in a gravitational field

A _____ B _____ C _____

D _____ E _____ F _____ (6)

2 Choose the option which best completes each of the following.

a) A unit used to measure energy could be a _____ (1)

newton **degree**

amp **joule**

b) A renewable energy resource that depends on naturally occurring valleys in mountainous regions is _____ (1)

wind **solar**

hydro **wave**

c) The store of energy possessed by a rollercoaster car waiting at the top of the ride is

_____ (1)

kinetic **gravitational potential**

electrical **thermal**

d) An example of a renewable energy resource is _____ (1)

a cell **oil**

biomass **coal**

e) Energy in a stretched spring is stored as a _____ (1)

gravitational store of energy **elastic strain store of energy**

thermal store of energy **light**

f) _____ is the energy resource that depends least on the Sun. (1)

Biomass **Hydro**

Wind **Tidal**

g) _____ is not a fossil fuel. (1)

Biomass **Natural gas**

Oil **Coal**

3   The diagram shows a small waterfall in a garden pond.

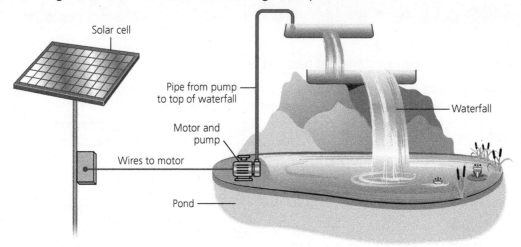

a)   Use terms from this list to describe how this waterfall works. The terms can be used more than once.

nuclear   thermal   gravitational potential   electrical   light   kinetic   sound

i)   _____ hits the solar cell and transfers energy from the _____

and _____ stores of energy in the Sun and reacts with the silicon crystals,

producing an electrical current.   (1)

ii) Energy is transferred through the wires via an _____ pathway to the

motor. Here the energy is stored as a _____ store of energy.   (1)

iii) As water flows from the top level to the pond, energy is transferred from a

_____ store of energy to a _____ store of energy.   (1)

b)   State **one advantage** and **one disadvantage** of using a solar cell to provide an energy resource to power the waterfall.   (2)

4   a)   Copy the words in the boxes below and then draw lines to match the fossil fuels to their common uses.   (4)

Fossil fuels	**Common uses**
Coal	Aircraft fuel
Natural gas	Generating electricity in power stations
Petrol	Heating and cooking in homes
Kerosene	Fuel for cars

b)   Fossil fuels are often described as 'non-renewable'. Explain what this means.   (1)

c)   i)   Much of the world's population uses biomass as a fuel. Explain what is meant by biomass fuel.   (1)

ii) Fossil fuels and biomass are both energy resources. Name the original source of this energy.   (1)

iii) Suggest **one advantage** of using fossil fuel rather than biomass as an energy resource.   (1)

5 An explorer has a wind-up torch. This torch is powered by a steel spring and does not use batteries.

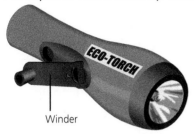

Winder

a) The explorer winds up the spring, and as the spring unwinds, energy is transferred to a small generator. The generator then turns and the bulb lights up.

   Copy and complete the following sentences to describe these energy transfers. (5)

   As the spring unwinds, energy is transferred from the _____ store of energy and turns the generator. The energy is now stored as a _____ store of energy in the generator. Light then transfers this energy to a store of _____ energy in the bulb. Some of the _____ store of energy is _____ to the surrounding air.

b) When the explorer turns up the brightness control, she notices that the spring unwinds more quickly. Explain why this happens. (1)

6 Choose the option which best completes each of the following.

a) The spreading out of energy from its store is called _____ (1)

   diffusion   **elimination**

   translation   **dissipation**

b) An electric drill is an example of a machine. When a drill is used, no

   _____ is emitted. (1)

   light   **sound**

   thermal energy   **kinetic energy**

c) The law of conservation of energy states that _____ (1)

   energy cannot be created or destroyed but can be transferred from one store to another

   energy input can never exceed energy output for a machine

   a machine is never 100% efficient

   energy transferred from chemical store of energy in a fuel can be transferred to many other stores of energy

7 The table below lists six methods of providing energy.

Method	How it works
Log stove	Burns dried wood to provide heat
Wave turbine	Uses waves to turn turbine which transfers energy via electricity (or via an electrical pathway)
Coal fire	Burns coal to release heat
Solar battery	Absorbs sunlight transferring energy from the nuclear store of energy to a chemical store of energy
Petrol generator	Burns petrol to generate electricity
Gas boiler	Burns gas to provide heat

a) i)   Name the energy resource for a solar battery. (1)

   ii) Explain how this same energy resource can drive a wind turbine. (2)

   iii) The energy resources used in solar batteries and wave turbines do not run out as they are used. What are these energy resources described as? (1)

b) Name **two** fossil fuels listed in the table. (2)

c) Copy and complete these sentences. (5)

Biomass fuels (e.g. _____) are made because plants carry out _____

which transfers energy into _____ stores of energy.

A _____ is a device that can transfer energy from one store of energy to another

to perform work. When this occurs, some energy is always transferred to the environment as

_____.

8 **a)** This diagram shows how much heat is transferred to the environment from different parts of a house.

Energy transfer = 10 000 J per minute

25% of the thermal energy that escapes goes through the roof.

10% of the thermal energy that escapes goes through the windows.

35% of the thermal energy that escapes goes through the walls.

15% of the thermal energy that escapes goes in draughts.

15% of the thermal energy that escapes goes through the floor.

i) State which part of the house transfers most heat to the environment. (1)

ii) Cavity wall insulation can reduce heat transfer through walls by 75%.

If the house had the walls insulated in this way, calculate the total transfer per minute from the insulated house. Show your working. (3)

iii) Copy and complete these sentences. (3)

Foam injected into the space between walls reduces the loss of heat because the foam

is a poor _____ of heat. Loft insulation works in a similar way, because the

_____ trapped between strands of fibreglass does not allow heat to be transferred.

Carpets stop heat escaping though the floor and also help to absorb _____ .

b) The diagram shows a simple draught excluder.

Efficient draught excluders can be expensive to fit. Explain how a draught excluder can nevertheless help to save money. (2)

Door

Gap

Excluder

9 Wave (tidal) energy can be used to generate electricity. One possible method is shown in this diagram.

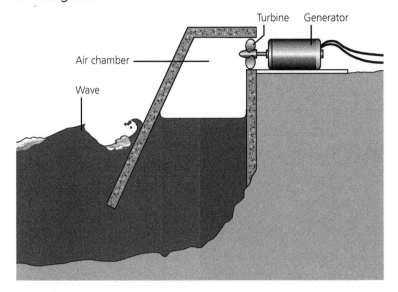

a) Each row in the table below shows a stage in generating electricity.

A	The turbine turns the generator
B	The moving air turns the blades of the turbine
C	The generator transfers energy as electricity
D	The waves move up the chamber
E	Air is pushed up the chamber by the waves

Arrange these stages in the correct order. (2)

b) A group of engineers investigated how the output of energy from the wave generator depended on the speed of the waves. The results are shown in the table below.

Wave speed, in m/s	0	5	10	15	20	25	30
Energy output, in kJ	0	0	7	25	50	78	105

These data were plotted on a grid.

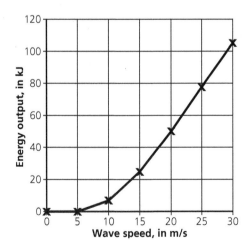

Suggest why there is no energy output if the wave speed is 5 m/s or less. (1)

17 Energy and electricity

1 Choose the option which best completes each of the following.

a) In a power station a jet of steam is used to turn the blades in a _____ (1)

furnace   **turbine**

boiler   **generator**

b) A mobile source of electricity is a _____ (1)

turbine   **conductor**

battery   **plug**

c) An example of a material that is not a good electrical insulator is _____ (1)

glass   **wood**

plastic   **graphite**

d) A _____ works by using electricity to transfer energy to a thermal store of energy. Light is emitted. (1)

solar panel   **battery**

TV monitor   **food mixer**

e) Energy in a kinetic store of energy can be transferred to another store of energy by electricity using a _____ (1)

battery   **generator**

petrol engine   **bicycle tyre**

f) Energy in a chemical store of energy is transferred to another store of energy by electricity using a _____ (1)

wind turbine   **generator**

battery   **microphone**

g) A lamp uses electricity to transfer energy to a _____ (1)

chemical store of energy and emits light

thermal store of energy and emits light

thermal store of energy and sound is heard

chemical store of energy and a thermal store of energy

h) A unit used to measure energy could be a _____ (1)

newton   **degree**

amp   **joule**

2 Electricity is a useful way of transferring energy from one store to another.

a) Give **two advantages** and **two disadvantages** of electricity as an energy pathway. (2)

b) i) The diagram below shows a simple electrical circuit.

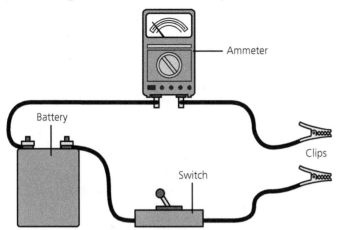

Explain how the apparatus could be used to compare the effectiveness of four different materials as insulators. **(2)**

ii) Explain **one** way in which you could make sure that this is a fair test. **(1)**

iii) Name **one** material that would be a good insulator, and describe **one** important use of this material. **(2)**

3   The diagram shows solar panels attached to the body of a satellite.

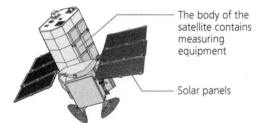

a) Scientists measured the output from one of these panels during one 24-hour period. Their results are shown in the graph below.

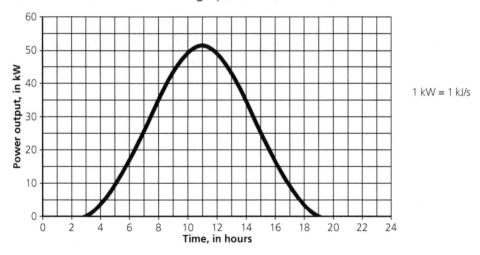

1 kW = 1 kJ/s

i) Explain why the power output varied during the 24-hour period. **(2)**

ii) The satellite used the solar panel to drive a motor. The motor needs 35 kW to run at full speed. Use the graph to work out how long the motor would be able to run at full speed. **(1)**

iii) The scientists decided to improve the design so that the solar panel turns to always face the Sun. Copy the graph and draw another curve to show how the power output for the new, improved solar panel would vary during the 24-hour period. **(2)**

b) Explain why solar panels are so useful on satellites and space stations. **(1)**

4   The diagram below shows the operation of a power station.

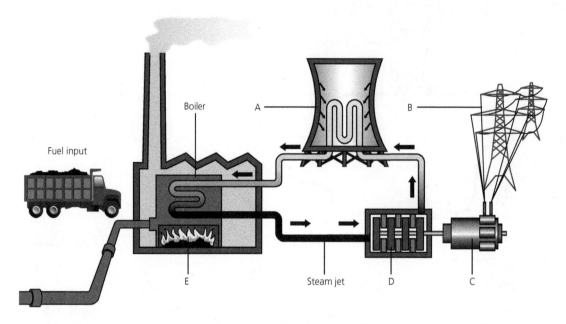

a) Choose words to complete the labels (A–E) on the diagram – not all of the words need to be used, and some could be used more than once.   (5)

furnace   water flow

generator   turbine

solar panel   light for power station

cooling tower   electricity output

b) The efficiency of a power station describes how much energy is transferred from a certain mass of fuel. Efficiency is measured in units of gigajoules per tonne.

   i)   The table below compares different types of power station. Work out the efficiency of each type of power station, giving your answers to the nearest whole number.   (4)

Type of fuel used	Fuel input, in tonnes	Energy output, in gigajoules	Efficiency, in gigajoules per tonne
Coal	1000	39 000	
Oil	2000	72 000	
Gas	1500	76 000	
Nuclear	500	21 000	

   ii) Give **two** reasons why we should use less coal in power stations.   (2)

   iii) Wind power is an alternative to the fuel types in the table above. Give **one advantage** and **one disadvantage** of wind power.   (2)

5   A classic car might use a dynamo as a generator. As the engine runs, it uses a pulley to turn the generator. The lights on the car are directly connected to the dynamo.

a)   Copy and complete the following sentences.   (5)

As the engine runs, the _____ store of energy in the petrol is transferred

into a _____ store of energy in the dynamo. This energy is transferred by

_____ carried in the wires to the lamps. The energy carried by the wires is transferred

to a thermal store of energy in the bulbs and _____ is emitted. Some of this

_____ store of energy is dissipated to the environment and is wasted.

b)   Modern cars have a battery between the dynamo and the lights. Explain why
this is a better arrangement for the car driver.   (2)

6   An electrical appliance uses electricity to transfer energy into stores of energy that are useful to us. Other energy pathways transfer energy to further useful stores of energy. The diagrams below show what happens in four appliances.

sound emitted	40%	
thermal store of energy	18%	
kinetic store of energy (to move whisk)	42%	

Food mixer

light emitted	12%
_____ (dissipated)	88%

Light bulb

sound emitted	_____
light emitted	40%
thermal store of energy	25%

TV monitor

sound emitted	18%
thermal store of energy	50%
kinetic store of energy	32%

Hair dryer

a)   i)   Calculate the percentage of energy dissipated by the food mixer   (1)

ii)   Calculate the percentage of the energy transferred by sound in the TV.   (1)

iii) Much of the store of energy in the light bulb is not useful. What is this store of energy?   (1)

b) Two pupils were interested in saving energy in school, and decided to investigate whether all energy-saving light bulbs were equally efficient. They obtained five different light bulbs, all rated as 9 watts, (which means they transfer the same energy per second), and they set up their apparatus as shown in the diagram below.

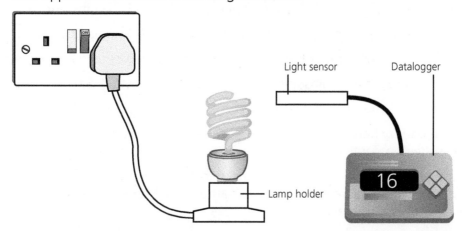

Each bulb was allowed to warm up for the same length of time, and the light sensor was always kept the same distance from the bulb. The pupils repeated each reading three times, and checked each other's readings on the datalogger. They obtained the following results:

Brand of bulb	Reading 1	Reading 2	Reading 3	Mean reading
Lumo	16	18	18	
Glo-bright	16	16	16	
Eco-save	18	19	17	
Brite-lite	21	19	20	
Supa-glow	18	18	19	

i) Calculate the mean value of light intensity for each bulb. (1)

ii) Plot these results of mean readings as a bar chart on a grid like the one shown here. (3)

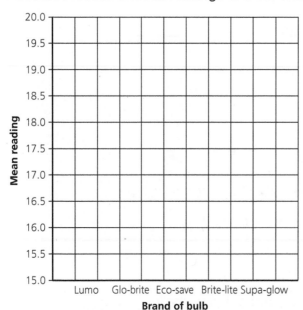

iii) Name the **independent variable** in their investigation. (1)

iv) Name the **dependent variable** in their investigation. (1)

v) Name **three fixed variables** that helped to make this investigation a fair test. (3)

vi) Explain how the pupils made sure their test was fair and their results were valid. (1)

vii) Give **one** conclusion that the pupils could have drawn from their investigation. (1)

18 Forces and linear motion

1 Choose the option which best completes each of the following.

a) Two cars are travelling in the same direction on a motorway. The one in the middle lane is travelling at 80 km/h and the one in the outside lane at 94 km/h. Their relative speed is

_____ (1)

$\frac{80}{94}$ km/h   −14 km/h

$\frac{14}{80}$ km/h $\frac{94}{80}$ km/h

b) Speed can be calculated from the formula _____ (1)

speed = time × distance   speed = $\frac{time}{distance}$

speed = $\frac{distance}{time^2}$ speed = $\frac{distance}{time}$

c) The unit in which force is measured is the _____ (1)

joule   watt

kilogram   newton

d) A motorcycle is being driven by a force of 100 N. The wind resistance is 28 N. The resultant

force moving the motorcycle is _____ (1)

$\frac{100}{28}$ N   −72 N

$\frac{28}{100}$ N   72 N

e) Forces always have _____ (1)

size only   a size and a direction

direction only   no value when an object is not moving

2   A drag racing car was being tested over a distance of 500 m. The digital stopwatch used to measure the time taken can measure to 0.01 seconds. The car was tested six times (three times in each direction).

Powerful engine

Light alloy wheels

a)   The results of the timing are shown in the table below.

Run number	Time taken, in s
1	6.02
2	6.23
3	6.00
4	6.19
5	8.24
6	6.21

   i)   Suggest which run was an anomalous result.   (1)

   ii) If the anomalous result is ignored, calculate the mean value for the time taken for the run. Show your working.   (1)

   iii) Explain how a mean value makes the results more valid.   (1)

b) i)   State the formula used to calculate the speed of the car.   (1)

   ii) Calculate the average speed of the car. Show your working, and give your answer to 1 decimal place.   (2)

c)   Give **one** possible reason why the results for runs 2, 4 and 6 were higher than those for runs 1 and 3, other than that the car travelled faster in one direction.   (1)

3   The diagram shows a submarine moving on the surface of the sea.

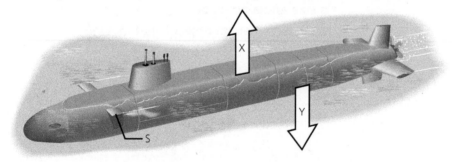

a)   Two of the forces acting on the submarine are marked X and Y. Identify these forces.   (2)

b)   The submarine now starts to move away from the dock. Name the **two** forces that resist the movement of the submarine when it is moving quickly.   (2)

c)   The commander of the submarine now decides that the vessel should dive below the surface. To do this, he must alter the angle of the stabilisers labelled S. Draw a sketch of the submarine and mark on the angle of the stabiliser as the submarine dives beneath the surface.   (1)

4   During the school cross-country race, the time taken by one of the runners to reach different distances around the course was measured. The results are shown in the graph below.

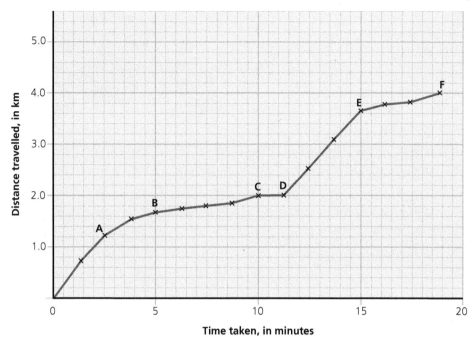

Time taken, in minutes

a) i)   State how long the runner took to complete the course.   (1)

    ii) Suggest at which point the runner had to stop to tie their shoelace.   (1)

    iii) State which part of the graph suggests that the runner was running up a steep hill.   (1)

    iv) Calculate the mean speed of the runner in km/h over the whole course. Give your
         answer to 1 decimal place. Show your working.   (2)

b) i)   Suggest why the runner wore trainers with ridges and spikes in the soles.   (1)

    ii) Copy and complete the following sentence.   (3)

         During the race, the runner's muscles transferred energy from a _____

         store of energy to a _____ store of energy and into a _____

         store of energy, which made them sweat.

5   Choose the option which best completes each of the following.

a) A metal which has a density of 4 g/cm³ and a mass of 30 g will have a volume of

    _____   (1)

    10.0 cm³   **30.0 cm³**

    15.0 cm³   **7.5 cm³**

b) The speed of a moving object will remain the same if _____   (1)

    it is in a vacuum   **all forces on it are balanced**

    there is no gravity acting on it   **there is a constant force on it**

6   Mark Marquez is a champion motorcyclist.
     His team works on his motorcycle to improve
     its performance by adding the fairing.

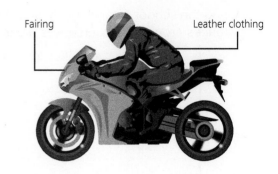

Fairing   Leather clothing

a)   i)   Explain how the fairing helps to improve
          the speed of the motorcycle.   (1)

     ii) The motorcycle does not slide off at the
          bends because of a force between the
          tyres and the track surface. State the
          name of this force.   (1)

     iii) If the motorcyclist falls from the machine, it is important that he slides along the
          track and slowly comes to a halt. Name the important property of his leather suit
          which helps this.   (1)

b)   Mark Marquez's engineers measured the distance travelled by his motorcycle along
     part of the straight. They obtained these results:

Time, in s	Distance travelled, in m
0	0
1	200
2	399
3	599
4	801
5	1000
6	1201
7	1400
8	1600

These results were plotted on a grid as shown below.

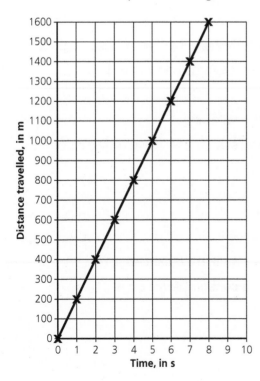

i) Calculate the average speed, in metres per second, during this period. (1)

ii) From the graph, determine how far the motorcycle had travelled after 3.5 s. (1)

iii) Determine how far the motorcycle travelled between 4.5 s and 7.5 s. (1)

iv) If the motorcycle continues at this speed, how long would it take to cover 2.5 km? (1)

7   Karen wants to investigate the relationship between mass and weight. She uses a forcemeter to do this and obtains the following results:

Mass, in g	Weight, in N
100	1.0
200	2.0
300	3.0
400	4.4
550	5.5
660	6.6

a) i) One of the results does not seem to fit the pattern of the other results. Suggest which is the anomalous result. (1)

ii) Use the information in this table to predict:

• the mass of an object weighing 4.3 N

• the weight of an object of mass 240 g. (2)

b) Copy and complete these sentences using words from the following list:

density   gravity   mass   shape   type   number

i) Weight is a force caused by _____ acting on an object. (1)

ii) _____ is not a force, and depends on the _____ and

_____ of particles in an object. (3)

c) An astronaut working on the International Space Station can take a spacewalk and can move using four small jet motors attached to their space suit. This diagram shows the size and direction of four forces acting on the astronaut.

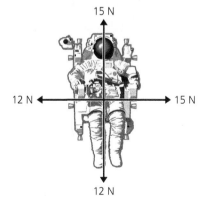

i) The astronaut can move even though the forces produced by the jets are small. Explain why. (2)

ii) Add an arrow to the picture of the astronaut to show the direction in which they would move. (1)

8   Rockets can be used to take astronauts into space. The diagram shows such a rocket shortly after take-off.

a)   The rocket is powered by burning hydrogen fuel. Tanks on the rocket carry hydrogen fuel and the oxygen needed to burn it.

   i)   Explain why the fuel and oxygen are transported as liquids rather than as gases.   (2)

   ii)   Explain why oxygen is needed to burn the fuel, although vehicles do not need liquid oxygen when they transport the rockets around the launch sites on Earth.   (1)

b)   i)   Name the **two** forces represented by the arrows alongside the rocket on the diagram.   (2)

   ii)   The graph below shows how the upward force and the weight of the rocket (including fuel) change during the first 40 seconds after ignition.

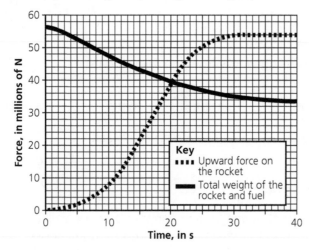

Use the graph to explain why the rocket cannot take off before 20 seconds have passed.   (1)

   iii) Calculate the resultant force on the rocket after 30 seconds. Show your working.   (2)

   iv)   Explain why the total weight of the rocket decreases during the first 30 seconds.   (2)

9   This diagram shows the effect of a mass M on a spring.

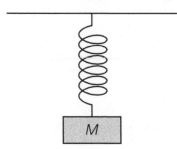

a)   Draw a diagram to show the effect of the same mass M on two identical springs joined in series (one after the other).   (1)

b)   Draw a diagram to show the effect of the same mass M on two identical springs combined in parallel (one alongside the other).   (1)

The suggested time for this test is 30 minutes.

1   Choose the option which best completes each of the following.

   a)   The unit for pressure is the _____   (1)

   newton   **newton-metre**

   newton/m²   **foot-pound**

   b)   Air resistance can also be called _____   (1)

   drag   **pressure**

   hindrance   **thrust**

2   A farmer tries to pull out a tree stump by pulling with a rope.

■ Not to scale

   a)   The tyres on the farmer's tractor have many grooves and ridges. Explain why this is
        important when working in a muddy field.   (1)

   b)   i)   Once the tree stump has been pulled out of the ground, the farmer decides to chop
             it up with an axe. The edge of the blade is sharpened.
             Explain why.   (1)

        ii) The blade of the axe has an area of 1.2 cm², and the farmer can apply a force of 600 N.
             Calculate the pressure the farmer can exert on the tree. Show your working.   (2)

3   Two pupils want to investigate friction between a surface and other materials. They use the
    apparatus shown in the diagram below.

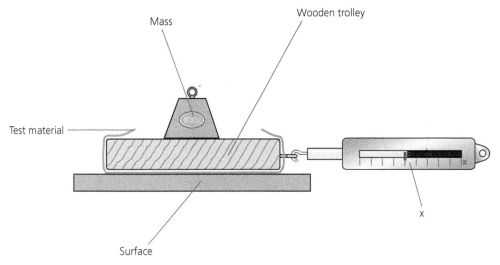

   a)   i)   State the name of the apparatus labelled X.   (1)

        ii) Explain why it might be necessary to add the mass to the trolley.   (1)

b) In this investigation, identify:

    i) the **independent variable** (1)

    ii) the **dependent variable** (1)

    iii) **two fixed variables**. (2)

c) Explain why the pupils should take three readings for each material. (1)

d) The pupils obtain the following results:

Material	Test 1, in N	Test 2, in N	Test 3, in N
P	2.0	2.0	2.0
Q	4.5	4.4	4.6
R	3.0	3.3	3.4
S	1.1	1.0	0.9
T	6.3	6.7	6.5

    i) Calculate the mean force in N for each of the materials. Give your answers to 1 decimal place. (2)

    ii) Suggest which material would be best for the soles of a rock climber's boots. (1)

    iii) Suggest which material would be best to rub onto the blades of a pair of ice skates. (1)

    iv) Suggest what would happen to the values for material R if the pupils polish the surface before they carry out the investigation. (1)

4 Scientists have observed that in some parts of the world polar bears are interbreeding with grizzly bears. The hybrid animal is called a grolar bear! Look at this image of a polar bear.

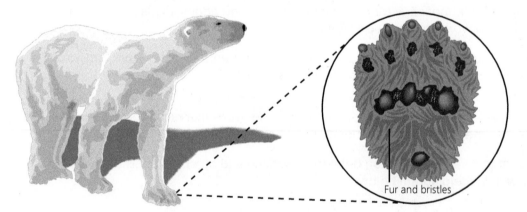

Fur and bristles

a) i) One difference between the two types of bear is that the true polar bear has much broader feet than the grizzly bear. Suggest how these broad feet could help the polar bear in its natural habitat, where it must walk across snowfields. (2)

    ii) There is also another difference in the feet of the polar bear and the grizzly. The polar bear has many hairs on the soles of its feet, but the grizzly does not. Some of these hairs are short (like bristles), and some are more like fur. Suggest the possible **advantages** to the polar bear, in its natural habitat, of the bristles and of the fur-like hairs. (2)

b) Copy and complete the following sentences. (2)

Inuit hunters who look for the polar bears rub fat and oil onto the runners of their

sledges. The oil acts as a _____ to reduce _____ between the

runners and the ice.

20 Sound

1 Choose the option which best completes each of the following.

a) The sound produced by a banjo string can be made louder by

_____ (1)

using a thinner string   **shortening the string**

plucking the string harder   **tightening the string**

b) Compared with light, sound travels _____ (1)

a little more slowly   **much more slowly**

much faster   **a little faster**

c) The diagram shows the structure of the human ear.

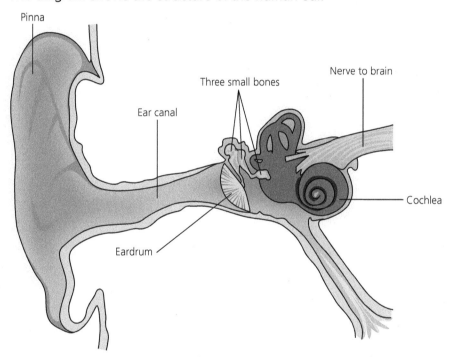

The first part of the human ear to vibrate when sound reaches it is the

_____ (1)

cochlea   **ear canal**

pinna   **eardrum**

d) Echo-sounding is not used to _____ (1)

detect submarines   **search for shoals of fish**

listen to a message on the phone   **locate sunken ships**

e) Sound _____ (1)

travels well through space   **only travels through air**

cannot travel through air   **cannot travel through a vacuum**

f) The size of a vibration is its _____ (1)

pitch   **amplitude**

frequency   **wavelength**

g) The frequency of a sound wave is measured in _____ (1)

joules   **millimetres**

hertz   **millivolts**

2   A teacher set up the piece of apparatus shown in the diagram.

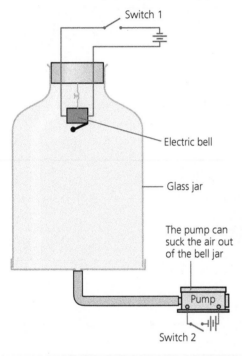

a) i)   Explain why you can hear the bell when switch 1 is closed. (1)

   ii) Describe what you can hear when switch 2 is closed. Explain your answer. (1)

b) Annie is not paying attention to the teacher – she is looking out of the window.
   She sees two gardeners hammering in new fence posts around the hockey field. Annie
   sees one of the men hit the post with a sledgehammer. One second later, she hears
   the sound.

   i)   Explain why she hears the sound after she sees the hammer hit the post. (1)

   ii) The gardener with the hammer moves halfway across the field, closer to the science
        laboratory. He starts to hammer on another post. How long is the gap between
        Annie seeing him hit the post and her hearing the bang? (1)

        there is no gap   **longer than one second**

        less than one second   **exactly one second**

3 a) The diagrams show the displays produced on an oscilloscope by four different sound waves.

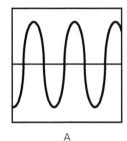 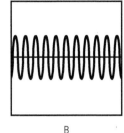 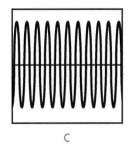 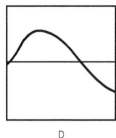

A   B   C   D

   i)   Identify which **two** sounds have the same pitch.   (1)

   ii) Identify which **two** sounds are as loud as one another.   (1)

b) i)   Builders using an electric drill might damage their ears. Explain how this could happen.   (2)

   The table below shows the maximum time, in hours, that a builder can work with a drill at different sound levels without damage to their ears. (Knowledge of the unit decibel is not required.)

Sound level, in decibels	Maximum time, in hours
86	8.0
88	4.0
90	2.0
92	1.0
94	0.5
96	0.25

   ii) Calculate the maximum time that a builder can work when drilling into concrete and making a sound of 89 decibels. Show your working.   (2)

4 The first astronauts landed on the Moon on 20th July 1969, wearing spacesuits.

   a) i)   The astronauts were able to talk to each other and to the spacecraft because they had radios, rather like mobile phones, in their helmets. Without the phones, they could not hear each other speaking. Explain why.   (1)

      ii) The astronauts were told that if their phones broke down, they could talk in an emergency as long as they touched their helmets together. Explain why they could hear each other when their helmets were touching.   (2)

   b) i)   The astronauts were instructed to recharge their radio batteries when they returned to the spacecraft. The main batteries in the spacecraft were recharged using a freely available renewable source of energy on the Moon. Suggest what this renewable energy source is.   (1)

      ii) Select which energy transfer takes place in the radio battery as it is being charged.   (1)

      chemical store of energy by sound

      thermal store of energy by electrical

      by sound to chemical store of energy

      by electricity to chemical store of energy

c) The astronauts could tell when they were being called because they could hear a buzz from the radio. The NASA technicians wanted to make sure that the buzz could be heard clearly.

The diagrams show the display on an oscilloscope for four possible sound waves.

The scientists thought that a loud sound with a high pitch would be best. Identify which of the oscilloscope displays matches this requirement. (1)

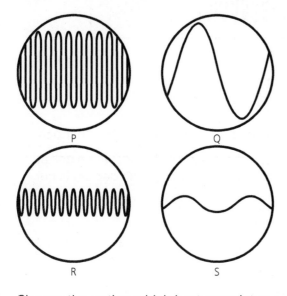

5 Choose the option which best completes each of the following.

   a) Permanent deafness can occur if _____ (1)

   the pinna is infected

   the nerve to the brain is damaged

   the eardrum is damaged by a pressure change

   the ear canal is blocked by wax

   b) A violinist plays a note followed by one of lesser frequency. The first note will have sounded _____ (1)

   softer   **lower**

   louder   **higher**

6 The starter at the school Sports Day uses a starting gun with blank cartridges. She stands 5 m away from the starting line for each race.

   a) The starter wears ear defenders. Explain why she does this. (1)

   b) Some senior pupils at the school were helping with the timing of the 200 m race. Each pupil was asked to record the time taken by a runner in a particular lane. One of them did not listen carefully to instructions, and started his stopwatch when he heard the bang from the gun and not when he saw the flash. Explain why the time he awarded to his runner would not have fitted in with the times awarded by the other pupil timers. (2)

c) The diagram below shows where the running track was located. Some of the spectators in the seats labelled S thought that they heard two shots at the start.

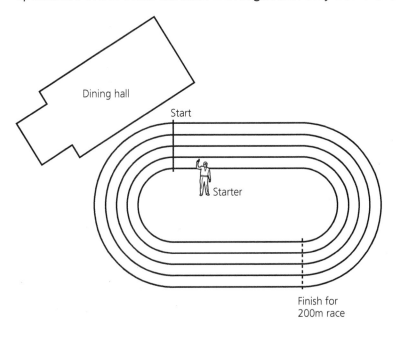

Explain how this could have happened. (2)

7 a) The nerve in the ear carries messages about sound to the brain. To make this possible, the structures in the ear transfer information by changing _____ (1)

light into electrical messages   **electricity into sound**

light into sound   **vibrations into electrical messages**

b) The table below lists the range of frequencies that six different animals can hear.

Animal	Lowest frequency, in Hz	Highest frequency, in Hz
Human	30	20 000
Robin	300	30 000
Dog	30	45 000
Cat	20	65 000
Dolphin	80	125 000
Bat	2000	110 000

i) A device sold by a pet shop claims to be able to frighten cats but not to affect dogs or birds, such as robins. It does this by emitting a loud, high-pitched sound when a cat passes in front of it. Suggest and explain a suitable frequency for the sound emitted by the device. (2)

ii) Dolphins find fish to eat by releasing short bursts of high-frequency sound. They work out how far away their prey is by measuring the time taken for an of the burst of sound to come back to them.

A dolphin emits a short sound burst and hears an echo 0.2 seconds later. Sound travels at about 1500 m/s in salt water. Calculate the distance of the prey from the dolphin (show your working and include the correct units). (3)

8   David wants to investigate some materials to find the best sound insulator. He places an electric bell inside a box, and then covers the box with each of the test materials in turn. The sound is detected by a sound sensor and recorded by a datalogger, as shown in the diagram below.

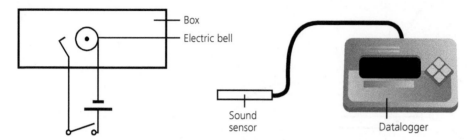

The results that David obtains are shown in the table below.

Insulating material	Sound level recorded, in units
None	68
Paper	60
Polystyrene block	35
Cardboard	54
Cloth	48

a) i)   Draw a bar chart to represent these results. Use a graph grid like the one below.   (4)

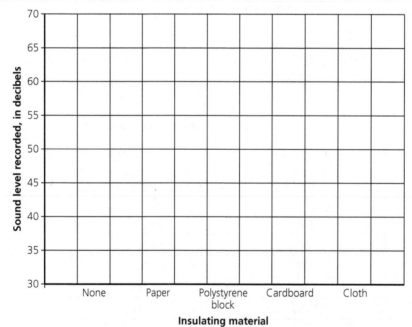

ii)   Identify which of the materials was the best sound insulator.   (1)

b) i)   For this investigation, name the **independent variable**.   (1)

ii)   Name the **dependent variable**.   (1)

iii)   Suggest which of the following are **fixed variables** in his investigation.   (2)

   - the box used
   - the distance between the sound sensor and the box
   - the person who records the results in the table
   - the light intensity in the room
   - the time of the day when the results are obtained

21 Light

1 Choose the option which best completes each of the following.

a) An example of a reflector and not a luminous source is _____ (1)

the Sun   **a motorbike's headlights**

the Moon   **a Bunsen burner flame**

b) A material that does not allow light to pass through it is _____ (1)

solid   **transparent**

opaque   **a shadow**

c) Compared with a Formula 1 racing car, light travels _____ (1)

more slowly   **much more quickly**

at about the same speed   **slightly more quickly**

d) Beams of light travel _____ (1)

at the same speed in every medium   **in straight lines**

at a speed that cannot be measured   **as a single colour**

e) The image observed in a plane mirror is not _____ (1)

upright   **back to front**

the same size as the object   **in front of the mirror**

f) When light is reflected from a plane mirror, the angle of _____ (1)

incidence is greater than the angle of reflection

reflection is greater than the angle of incidence

incidence is equal to the angle of reflection

incidence is less than the angle of reflection

g) The bouncing of light rays from a surface is _____ (1)

dispersion   **reflection**

refraction   **diffraction**

h) Different colours of light are made of waves with _____ (1)

different frequencies   **different densities**

different amplitudes   **different angles**

i) The splitting of light by a prism is _____ (1)

dispersion   **reflection**

refraction   **diffraction**

j) A rainbow is caused by _____ (1)

light from our eyes being refracted by raindrops

sunlight reflected from the surface of water drops

sunlight being dispersed as it passes through water drops

sunlight reflected from the sky

2 Two students are interested in how shadows are formed and use the apparatus shown in this diagram.

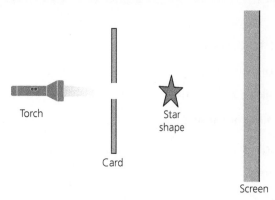

a) Explain why the card and the star shape must be opaque. (1)

b) Draw **two** lines representing light rays to show how the shadow of the star is formed. (2)

c) Describe what would happen to the size of the shadow of the star on the screen if the shape was moved further away from the card. (1)

d) State the word used to describe a material that allows some light to pass through it. (1)

3 Two mirrors held at 90° to each other always reflect a ray of light parallel to the incident ray.

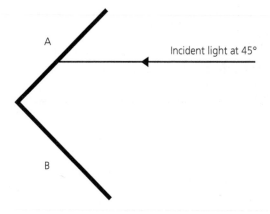

a) In the diagram above, a ray of light strikes mirror A at an angle of 45°. Copy the diagram and then use a ruler and protractor to complete the diagram to show how the mirrors reflect the ray. (2)

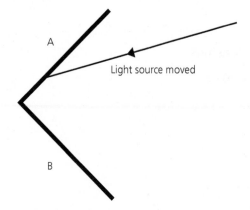

b) In the diagram above, the source of light has been moved so that the ray of light strikes mirror A at a different angle. Copy the diagram and then use a ruler and protractor to complete the diagram to show how the mirrors reflect the ray. (3)

c) Police officers, ambulance workers and firefighters wear reflective jackets when working. The reflective stripes on the jackets are made up as shown in the diagram below.

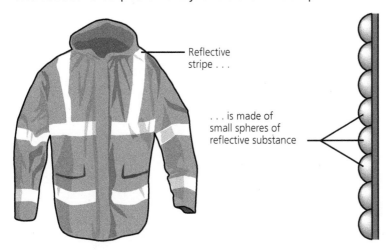

i) At night, road users can see these reflective jackets in the beam of their headlights. Explain why. (2)

ii) Explain why a plane mirror would not be suitable for a reflective jacket. (1)

iii) Explain why the cloth trousers of a firefighter (which do not have reflective stripes) do not give a clear reflection. (1)

4 The diagram below shows a ray of white light passing through a perspex block.

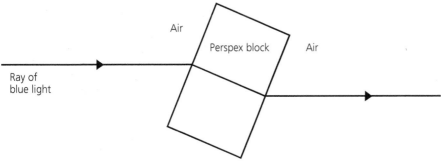

a) As the light goes into the block it changes direction. State the name of this effect. (1)

b) The light leaving the block is not as bright as the light entering the block. Explain why. (1)

c) Light can be made to pass through a prism, and to form a spectrum on a white screen.

Copy and complete this sequence to describe the colours of the spectrum in the correct order. (4)

_____ _____ yellow _____ _____

_____ violet

5   At a zoo, a rare bird has made a nest. The keepers at the zoo do not want to disturb the bird but want to allow visitors to view the nest. They build a piece of equipment like the one shown below.

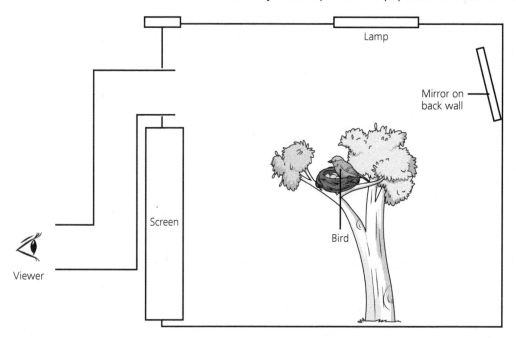

a)   Copy the diagram and add mirrors to show how the visitors will view the nest.   (2)

b)   Draw a ray of light from the bird to show how it reaches the visitor's eye. Include arrows on the ray to show its direction.   (1)

c)   Observers using binoculars for birdwatching hope to see a clear image. The most expensive binoculars have their lenses coated with a special material. The makers of the binoculars claim that this coating 'reduces dispersion'.

     Explain why these expensive binoculars provide a more accurate coloured image than cheaper, non-coated, binoculars do.   (2)

6 A teacher uses a 'light pointer' to point at diagrams on a whiteboard. The pointer gives a very fine beam of red light.

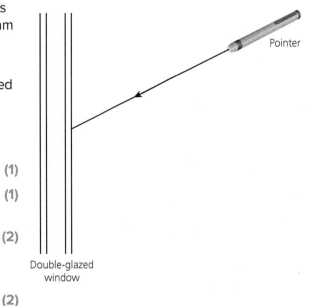

If the teacher points the beam away, at a double-glazed window, the students can see two spots of red light.

a) Copy this diagram, then on the diagram:

  i) Draw the beam of light reflected off the first glass surface. **(1)**

  ii) Label the angle of reflection. **(1)**

  iii) Show how the beam is reflected to give the second red spot. **(2)**

b) The beams are not as bright when they are reflected from the window as they are when reflected from a mirror. Explain why. **(2)**

7 A snooker player is ready to play a shot, hitting the white ball against the blue. The table is well lit by a source of white light.

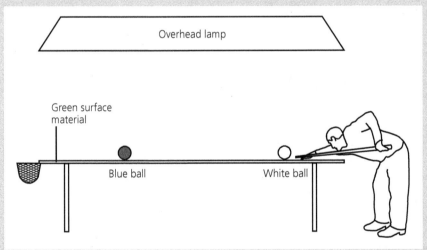

a) Describe how light from the overhead lamp lights up the balls and makes them visible to the player. **(2)**

b) State which energy transfer takes place in the eyes to allow our brain to 'see' an object. **(1)**

c) If the overhead lamp stops working, the players might use a light source from the side of the table. Explain why this will cause shadows behind the snooker balls. **(1)**

d) Light from the overhead lamp shines onto the snooker balls and onto the material of the table surface. Explain how the reflection from the balls is different from the scattering of light by the material of the table. **(2)**

22 Electrical circuits

1 Choose the option which best completes each of the following.

a) Energy in a cell, stored as _____ (1)

a chemical store of energy, is transferred by light

a chemical store of energy, is transferred by electricity

a kinetic store of energy, is transferred by electricity

a kinetic store of energy, is transferred light

b) A chemical store of energy which can transfer energy via an electrical pathway

is a _____ (1)

terminal   **cell**

fuse   **transistor**

c) A device that allows the flow of current in a circuit is a _____ (1)

resistor   **terminal**

switch   **pole**

d) The unit of electrical current is the _____ (1)

amp   **hertz**

joule   **newton**

e) A material that allows electricity to pass through it is _____ (1)

an insulator   **a circuit**

a cell   **a conductor**

f) A circuit with all of the components joined in a single loop is a

_____ (1)

parallel circuit   **conducting circuit**

series circuit   **short circuit**

g) A component that can control the flow of current is _____ (1)

a resistor   **an ammeter**

a cell   **a motor**

h) Adding an extra cell to a series circuit that includes a lamp will make the lamp

_____ (1)

shine less brightly   **shine more brightly**

shine as brightly as before   **go cooler**

i) A circuit in which the current can only take one pathway is a _____ (1)

parallel circuit   **switched circuit**

series circuit   **short circuit**

2 a) Link the correct name to each circuit symbol. Note that there are more names than symbols. (5)

Circuit Symbol
A —⌒o—
B —⊗—
C —|⊦⸱⸱|⊦—
D —(A)—
E —▭—

Name
Ammeter
Fuse
Switch
Motor
Battery
Lamp
Resistor

b) James made the circuit shown in the diagram.

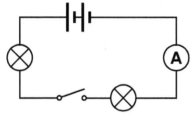

i)   Name the energy store for the circuit. (1)

ii) State which component in the circuit is used to measure the current flow. (1)

iii) Redraw the circuit so that each bulb shines more brightly. (2)

c) Name the most commonly used metal to make the wires in electric circuits. (1)

3   The diagram shows a room heater used to warm up the school gym on cold mornings.

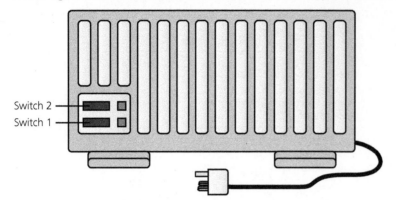

The school electrician had a diagram that showed the circuit for the heater.

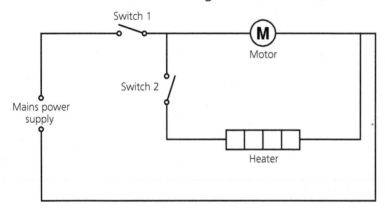

a) i)   State which switches must be closed for the heater to work. (1)

   ii)   Explain whether it is possible to have the heater on when the blower motor is
         switched off. (1)

   iii) The heater and the motor are both on. A wire in the heater breaks. Suggest what
         effect this will have on the blower motor. (1)

b) Copy and complete this table to compare series and parallel circuits. (4)

Feature	Series circuit	Parallel circuit
Value of current in different places		
Number of pathways that current can take		
Effect of one damaged component		

4 **a)** The diagram below shows the parts of a cycle lamp.

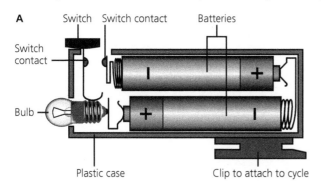

i) Anna closes the switch. Explain why the lamp lights. (1)

ii) Use the correct symbols to draw a circuit diagram for the lamp. (3)

b) Jamie borrowed the cycle lamp. The lamp would not light, even when the switch was closed. The two diagrams show possible reasons for this.

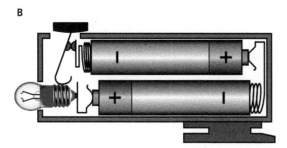 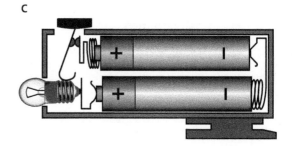

In each case, B and C, describe and explain what needs to be done to get the lamp to light. (4)

5   Janet and Imran are asked whether the thickness or the length of a piece of wire is more important in affecting its resistance. They are supplied with the following equipment:

- battery
- roll of thin copper wire
- switch
- roll of thin steel wire
- ammeter
- ruler with millimetre markings
- roll of thick copper wire

a) Draw a circuit that they could use in their investigation.   (2)

b) For their investigation to be a fair test, identify the:

   i)   **independent variable**   (1)

   ii) **dependent variable**   (1)

   iii) **fixed variables**.   (2)

c) Describe how they could try to make sure that their results are valid.   (1)

6   A teacher set up the circuit shown in this diagram.

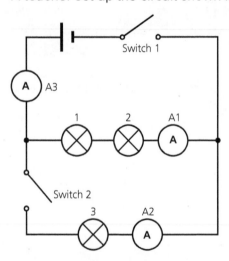

a) A current of 2.2 A flows through ammeter A3 when switch 1 and switch 2 are closed.

   i)   Describe what will happen to each of the lamps when switch 1 is closed with switch 2 open. (1)

   ii) If ammeter A2 has a reading of 1.0 A when switch 1 and switch 2 are closed, state what the reading on ammeter A1 will be.   (1)

   iii) State what the reading on ammeter A1 will be if switch 1 is closed and switch 2 is open.   (1)

b) Copy and complete the paragraph about electric circuits. Use words from this list. There are more words than spaces to be filled. (4)

buzzer   lamp   cell   lead   conductor   resistor

current   socket   filament   switch   insulator

Electricity can pass through any material that is a _____. A complete circuit

lets the _____ flow all the way round it. The energy can be supplied by a

_____ and can pass from one component to another through a _____.

When a circuit is made up, it may include a _____ which can be opened to

stop the flow of current. If the _____ is closed, then a component such as a

_____ will light or a _____ will sound.

7 a) A teacher is demonstrating circuits to her pupils. She builds this circuit.

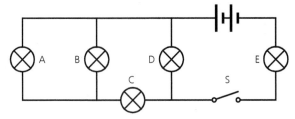

She closes the switch and all of the lamps come on. One of the lamps then fails and all of the lamps go out. Identify which one of the lamps must have failed. (1)

b) The teacher then builds a second circuit. She includs a metal pencil sharpener and an eraser in different parts of the circuit.

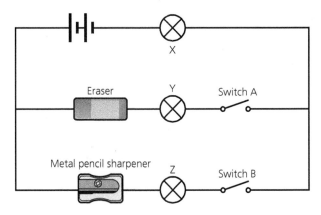

Copy and complete the table below to show which lamps in this circuit will be on and which will be off when the switches are open or closed. (2)

Switch A	Switch B	Lamp X	Lamp Y	Lamp Z
Open	Open	Off	Off	Off
Closed	Open			
Open	Closed			

c) Finally, the teacher builds this circuit, using a battery, three lamps and four ammeters.

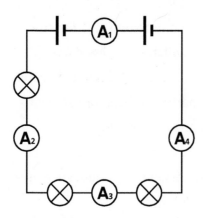

The current reading at ammeter A_1 is 0.6 A. State which set of readings for the other ammeters, shown in the table below, is correct.

(1)

Reading on A_2, in A	Reading on A_3, in A	Reading on A_4, in A
0.2	0.2	0.2
0.3	0.0	0.3
0.6	0.6	0.6

23 Magnets and electromagnets

1 Choose the option which best completes each of the following.

a) The ends of a magnet are called _____ (1)

tips   **terminals**

poles   **fields**

b) An example of a metal that has magnetic properties is _____ (1)

iron   **copper**

tin   **aluminium**

c) The rolled wire making up part of an electromagnet is the _____ (1)

coil   **core**

turning   **relay**

d) A compass needle shows the direction of the magnetic North Pole because the compass
needle is _____ (1)

magnetised   **a good conductor**

unmagnetised   **heavier at one end**

e) The symbol for a relay is _____ (1)

A   B   C   D

f) A bar magnet has two poles, north and south;

_____ will attract a piece of unmagnetised iron.

both poles   **only the north pole**

neither pole   **only the south pole**

2   a)   David is investigating the properties of magnets and wants to find a magnetic material. He has three pieces of metal – one is made of steel, one is made of tin and one is a magnet. He does not know which piece is which, but marks each one with a letter (X, Y or Z) and then uses a bar magnet to try to identify them.

He places the marked end of each piece of metal next to each pole of the bar magnet and writes down what happens in a table of results. Copy and complete this table of results.   (3)

Test			Result	Conclusion
X	S	N	Attracts	Metal X is
X	N	S	Attracts	_____
Y	S	N	_____	Metal Y is
Y	N	S	Attracts	_____
Z	S	N	_____	Metal Z is
Z	N	S	Nothing happens	_____

b)   David takes the piece of metal that was magnetised and lays it beneath a piece of card. He then sprinkles iron filings onto the card. Draw a diagram showing the pattern he will see.   (2)

c)   David then takes a small compass and places it near to the magnet. Draw a diagram to show what happens to the compass.   (1)

d)   Copy and complete this paragraph.   (4)

When an unmagnetised iron nail is put into a _____ it becomes magnetised.

The south pole of this nail will be _____ to the _____ pole of a bar

magnet, but will be _____ by the north pole of the magnet.

3 A current flowing through a coiled wire acts like a magnet. The strength of this electromagnetic field can be increased by placing a core inside the coiled wire. A pupil decides to investigate the effect of the core material on the strength of the electromagnetic field. She uses the apparatus shown in the diagram below.

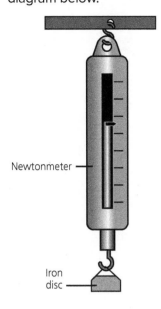

Newtonmeter

Iron disc

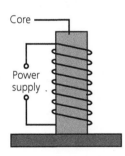

Core

Power supply

a) Name **three** factors that should be kept constant to keep this as a fair test. (3)

b) The pupil obtained the following results.

Material in core	Reading on newtonmeter, in N
No core	1.4
Iron (no current)	1.0
Iron	1.8
Glass	
Steel	1.6

   i) Explain why the reading on the newtonmeter increases when a current passes through the coil. (2)

   ii) Suggest the likely value for the reading with glass as the core. (1)

4   Andy makes two electromagnets, as shown below. The strength of the electromagnet is measured by how many paper clips can be picked up.

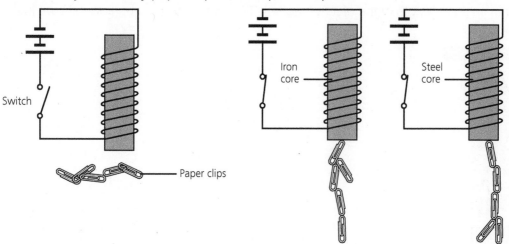

Switch

Paper clips

Iron core

Steel core

a) i)   Explain how you can tell that the strength of both electromagnets is the same.   **(1)**

    ii) When the switches are opened, the paper clips fall from the iron core but not from the steel core. Explain why iron, rather than steel, is used for the core of an electromagnet.   **(1)**

b) The diagram shows an electromagnet used in a scrapyard for separating different metals.

    Explain how the electromagnet can separate the valuable aluminium from the less valuable iron and steel.   **(2)**

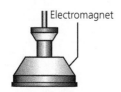

Electromagnet

Pile of metal:
mixed iron/steel/aluminium

c) Some parts of the electromagnet crane are protected by circuit breakers. These automatically switch off a circuit if the current is too high. This diagram shows a simple circuit breaker.

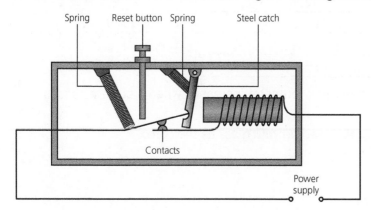

Spring   Reset button   Spring   Steel catch

Contacts

Power
supply

    i)   Explain why the current is cut off by this circuit breaker if the current is larger than a certain value.   **(2)**

    ii) Give **one advantage** of this type of circuit breaker compared with a simple electrical fuse.   **(1)**

5   Choose the option which best completes each of the following.

a)   A magnetic field has _____ (1)

force but no direction   **direction but no force**

both force and direction   **neither force nor direction**

b)   An electromagnet can be made stronger by increasing the _____ (1)

diameter of the coil   **length of time the current is passed**

insulation on the wire   **number of turns in the coil**

c)   _____ does not use an electromagnet. (1)

A relay   **An electric bell**

A compass   **A DC motor**

d)   The symbol for a reed switch is _____ (1)

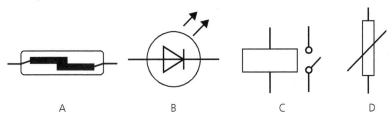

A   B   C   D

6   A reed switch is a small relay used in electronic circuits. It has thin metal contacts inside a glass tube.

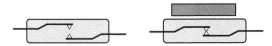

a)   Jane sets up the circuit shown below.

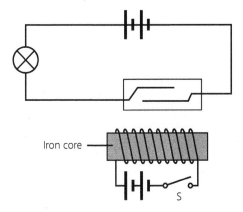

Iron core

i)   She closes switch S but the lamp does not light. She knows that all of the connections whave been made properly, and that the lamp is not broken. Explain why the lamp might not light up. (2)

ii) Suggest **two** things Jane could do to the electromagnet to overcome this problem. (2)

7 Saira uses a sensor to measure the strength of an electromagnet. She places the sensor 50 mm from the electromagnet and increases the current in the coil. She then turns the current down to zero, moves the sensor to 100 mm from the electromagnet and repeats the experiment.

The results are shown in the tables.

Sensor at 50 mm distance	
Current, in amps	Sensor reading, in N
0.5	0.35
1.0	0.68
1.5	1.00
2.0	1.30
2.5	1.50

Sensor at 100 mm distance	
Current, in amps	Sensor reading, in N
0.5	0.15
1.0	0.30
1.5	0.45
2.0	0.55
2.5	0.60

a) The diagram shows a graph of these results.

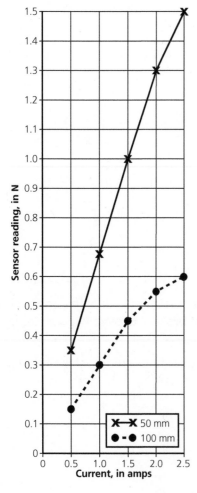

i)   State how the size of the current in the coil affects the strength of the electromagnet. (1)

ii) Suggest **two** other ways in which Saira could alter the strength of the electromagnet. (2)

b) An electromagnet can be used at a railway crossing barrier.

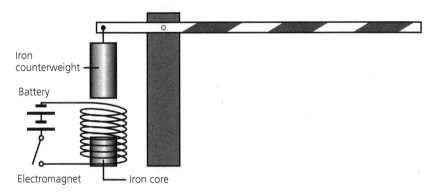

Explain how the electromagnet can be used to raise the barrier. **(2)**

8 The end of morning school is normally signalled by ringing an electric bell. The diagram shows the circuit for this bell.

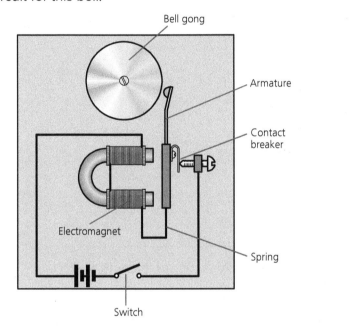

a) The bell is normally silent. Explain why. **(2)**

b) Here is a list of possible actions in this circuit:

- electromagnet pulls on armature
- spring pushes armature back again so circuit is completed once again
- switch is closed
- contact breaker breaks circuit
- armature moves so gong strikes bell

Write out, in the correct order, what happens to make the bell ring. **(5)**

24 Space

1 Choose the option which best completes each of the following.

a) On the Moon, gravity exerts a force of 1.6 N on 1 kg. An object that has a weight of 240 N on the Moon, has a mass of _____ (1)

400 kg   **160 kg**

150 kg   **240 kg**

b) An eclipse of the Moon occurs when _____ (1)

the Sun lies between the Earth and the Moon

the Moon lies between the Sun and the Earth

the Moon is waning

the Earth lies between the Sun and the Moon

c) The Sun is a _____ (1)

constellation   **galaxy**

star   **planet**

d) The time taken for the Moon to complete one orbit of the Earth is a

_____ (1)

year   **day**

lunar month   **season**

e) We are sometimes able to see planets because _____ (1)

they are luminous   **asteroids collide with the planet's surface**

they reflect light from the Moon   **they reflect light from the Sun**

f) _____ is the correct formula for the calculation of the weight of an object. (1)

$\dfrac{mass}{gravitational\ field\ strength}$ $mass^2 \times gravitational\ field\ strength$

$\dfrac{gravitational\ field\ strength}{mass}$ $mass \times gravitational\ field\ strength$

g) The planet with an orbit closest to the orbit of the Earth is _____ (1)

Uranus   **Jupiter**

Mercury   **Venus**

h) Compared with sound, light travels _____ (1)

much slower   **at the same speed**

much faster   **slightly faster**

i) The correct units for gravitational field strength are _____ (1)

kilograms per newton **newtons per gram**

newtons per kilogram **newtons per metre**

j) Distances in space are so great that they are measured in _____ (1)

millions of kilometres **sound years**

light years (distance travelled **billions of kilometres**
by light in a year)

2 Copy and complete the following sentences. (5)

We are able to see stars because they are _____. Many stars seem to be arranged

in patterns called constellations, and there are millions of stars in a single _____.

The stars may be a great distance from the Earth and may only be visible with an instrument

called a _____. We have to measure distances in _____. All the

planets, stars, gases and dust together make up the _____.

3 The diagram shows a satellite in orbit around the Earth.

Geostationary satellite

■ Not to scale

a) Copy the diagram and draw lines to show how a transmitter and receiver would allow a football match to be seen in a different country. Do not worry too much about drawing the countries accurately. (2)

b) This type of satellite moves at a speed and height that means it appears to remain in the same position relative to the surface of the Earth. It is called a geostationary satellite.

   i) Explain why it is important for the satellite to remain in the same position above the Earth. (1)

   ii) Once a TV satellite dish is fixed to the outside of a house, it does not need to be moved. Explain why. (1)

   iii) State how long one complete orbit of a geostationary satellite takes. (1)

   iv) Name the force that keeps the satellite in position above the Earth. (1)

c) Name **one** natural satellite of the Earth and **one** natural satellite of the Sun. (2)

4   Copy the words in the boxes below and then draw lines to match each observation to the correct explanation. (5)

Observation

1 year on Earth is 365 days

At the equator, there are 12 hours of light and 12 hours of darkness

In Britain there are four seasons in the year

There is a new Moon every month

A ship sailing away from land goes out of sight

Explanation

The Earth's axis is tilted

The Moon orbits the Earth

The Earth orbits the Sun

The Earth is a sphere

The Earth rotates on its axis

5   This diagram shows a model of the solar system.

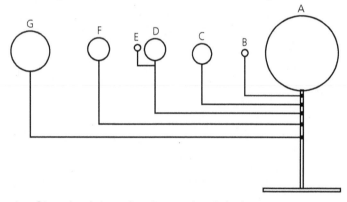

a) Give the letter that is used to label:

   i)   the model Earth (1)

   ii) the model planet that would be expected to have the highest surface temperature (1)

   iii) a star. (1)

b) Spacecraft have allowed humans to stand on the surface of the Moon. This diagram shows an astronaut standing at four different positions on the Moon.

   i)   Copy the diagram and draw an arrow at each of the four positions to show the direction of the force of the Moon's gravity on the astronaut. (1)

   ii) The astronaut is holding a bag for collecting samples on a chain. Draw the position of the bag in positions X, Y and Z. (1)

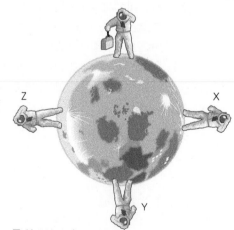

■ Not to scale

c) The diagram shows that the Earth orbits the Sun.

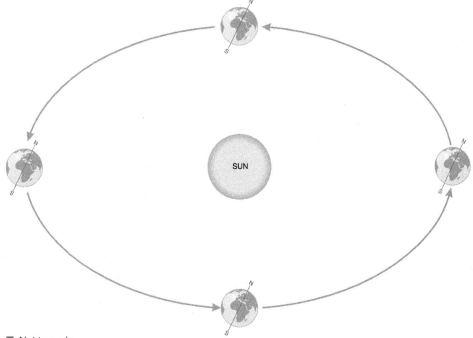

■ Not to scale

i) Explain why the Earth orbits the Sun. (2)

ii) State how long it takes for the Earth to orbit the Sun once. (1)

iii) Light travels at 300 000 km/s. The Sun is 149 million km from the Earth. Calculate how long light takes to reach the Earth from the Sun. Show your working. (2)

6 The diagram shows our solar system.

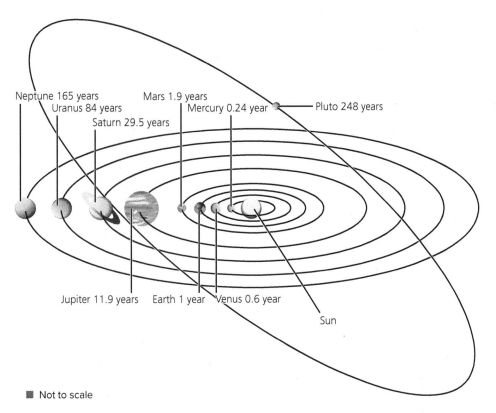

■ Not to scale

a) i) Explain how evidence from the diagram supports the idea that Pluto is not a planet. (1)

ii) The Hubble telescope has allowed astronomers to observe an object called Charon, which orbits Pluto. Suggest how this supports the idea that Pluto is a planet. (1)

7   The diagram below shows the positions of the Earth, Moon and Sun during a lunar eclipse.

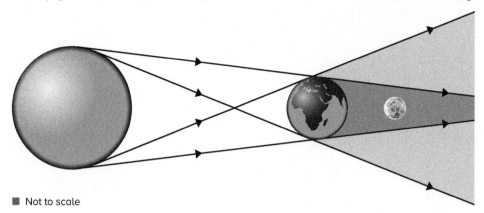

■ Not to scale

a)   Redraw the diagram and add labels to the Sun, Moon, Earth and partial shadow.   (2)

b)   Draw a second diagram showing the position of the Sun, Moon, Earth and complete shadow (umbra) during a solar eclipse. Label your diagram.   (3)

c)   A solar eclipse can upset the singing patterns of birds. Suggest why this is.   (1)

Exam Practice Answers

Experiments in science (page 7)

1 a) carbon dioxide   (1)

   b) white to blue   (1)

   c) oxygen   (1)

   d) deep blue   (1)

   e) an independent variable   (1)

   f) hydrogen   (1)

   g) a pipette   (1)

2 a) i) E   (1)

      ii) D   (1)

   b) i) wear safety glasses   (1)

      ii) stopwatch/digital timer   (1)

      iii) thermometer   (1)

3 a) i) air hole open or closed   (1)

      ii) time taken for water to boil   (1)

   b) the position of the Bunsen burner below the beaker; the volume of water in the
beaker; the position of the gas tap (i.e. how much gas flows)   (3)

4   (3)

Effect on limewater	pH with universal indicator	Effect on a burning splint	Gas
None	4	Puts it out	Sulfur dioxide
None	7	Goes 'pop'	Hydrogen
Turns it cloudy	6	Puts it out	Carbon dioxide
None	7	Burns more brightly	Oxygen

5 a) i) measuring cylinder (beaker is also acceptable, but is less accurate) (1)

      ii) mm-ruled ruler   (1)

   b) i)   ii)

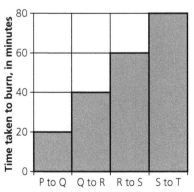 (1)

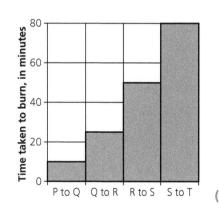 (1)

  c) any three of:

   • same thickness of candle wicks
   • same length of candle wicks
   • same wax
   • same dimensions
   • same distance between markings (3)

6 a) i) to make sure that the hot vapour condensed to a liquid (1)

    ii) carbon dioxide (1)

    iii) most obvious colourless liquid to test for is water – two possible tests: adding water to blue cobalt chloride paper will turn it pink; adding water to white anhydrous copper sulfate will turn it blue; these colour changes, in either cobalt chloride or copper sulfate, demonstrate that the colourless liquid is water (2)

  b) i) no naked flames; no drinking or eating (2)

    ii) the concentration of antifreeze is 10% and this protects only to $-5\,°C$; therefore, the water in the radiator will freeze and the radiator could split (water expands as it freezes) (2)

7 a) A – irritant; B – oxidising; C – corrosive; D – highly flammable; E – toxic (5)

  b) highly flammable; irritant; harmful to the environment (3)

8 a) mass of water = 196.4 g – 116.4 g = 80 g

    density of water = $\dfrac{80\,g}{80\,cm^3}$ = 1 g/cm³

    mass of alcohol = 180.2 g – 116.4 g = 63.8 g

    density of alcohol = $\dfrac{63.8\,g}{79\,cm^3}$ = 0.8 g/cm³ (3)

  b) pipette (1)

Biology
1 Cells and organisation (page 12)

1 a) transferring energy (1)

  b) sperm cell (1)

  c) producing offspring (1)

  d) taking in nutrients (1)

  e) a cell (1)

  f) a tissue (1)

  g) mitochondrion (1)

  h) methylene blue (1)

2 digestive – teeth and stomach

   circulatory – heart

   reproductive – testes

   breathing – lungs

   skeletal – rib (5)

3   a)   i)   D   (1)

         ii)   to contain genes that control the production of proteins in cells   (1)

     b)   B   (1)

     c)   C and E   (2)

     d)   E   (1)

4   a)   i)   chloroplast   (1)

         ii)   no photosynthesis takes place in roots   (1)

     b)   cell wall – helps to give the cell a definite shape

         cell membrane – controls the entry and exit of substances

         nucleus – contains the genetic material that controls the cell's activities

         cytoplasm – many chemical reactions take place here

         chloroplast – absorbs energy from the Sun for photosynthesis   (5)

5   a)   secrete mucus to trap microorganisms and particles of dust   (1)

     b)   root hair cell   (1)

6   a)   i)   chloroplast; cell wall   (2)

         ii)   chloroplast (photosynthesis – to provide sugar for the plant);

               cell wall (to help the cell keep its shape / to prevent the plant cell bursting)   (2)

         iii) A – cell membrane (to control what enters and leaves the cell);

               B – cytoplasm (place where chemical reactions occur inside the cell)   (4)

     b)   i)   cells develop certain features that suit them to one particular function   (1)

         ii)   red blood cell – transports oxygen – respiration

               leaf cell – traps light energy – photosynthesis

               root hair cell – absorbs mineral ions – plant nutrition

               egg cell – carries genes from female – reproduction   (4)

7   a)   A   (2)

     b)   cell wall; cilium; chloroplast   (3)

2 Nutrition and digestion (page 16)

1   a)   protein   (1)

     b)   fibre   (1)

     c)   starch   (1)

     d)   help develop strong bones   (1)

     e)   protein   (1)

     f)   incisors   (1)

     g)   sugar   (1)

     h)   canines   (1)

2   a)   too much salt – high blood pressure

         too little iron – cannot carry enough oxygen in blood

         too much fat – heart disease

         not enough fibre – constipation

         too little protein – slow growth of muscles   (3)

b) sugar – the main source of energy for working cells

calcium – required for development of bones and teeth

vitamin C – prevents scurvy

water – an important part of the process of digesting foods

starch – provides a supply of sugar (3)

3 a) i) carbohydrate (1)

   ii) fat (1)

   iii) cheese (1)

   b) i) 12 g (1)

   ii) water (1)

   c) i) 1000 g or 1 kg (1)

   ii) yes (1 kg of brown rice provides 95 g of protein) (1)

   d) i) 1000 mg (less than a pregnant woman requires, but more than a non-pregnant woman in order to provide calcium for the growing baby) (1)

   ii) she is still growing and producing bone and teeth (1)

   iii) to produce materials for growth and repair of tissues (1)

4 calcium (1)

5 a) i) less fat – reduced chance of heart disease/obesity

less cholesterol – reduced chance of heart disease

more fibre – reduced chance of colon cancer (3)

   ii) more protein for growth

more energy for exercise (2)

   b) i) USA (1)

   ii) four times more likely (= $\frac{20}{5}$) (2)

   iii) wholemeal bread; apples (1)

6 a) (3)

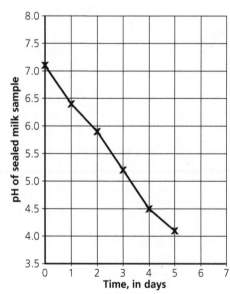

   b) lactic acid (1)

   c) 3.2 (accept values of 3.1–3.3) (1)

d) allow marks for any reasonable method, provided that the following variables are identified correctly:

- temperature – independent variable
- pH – dependent variable
- time for heating and amount of milk – controlled variables (4)

e) yoghurt or cheese (1)

f) keep the milk at a low temperature – bacteria will not multiply so quickly

dry the milk and store it as granules – bacteria cannot multiply without water (2)

7 a) i) iodine solution (1)

ii) Benedict's solution/reagent (1)

b) this is human body temperature/the optimum temperature for amylase activity (1)

c) i) no starch can cross from inside the bag to the water in the beaker (because starch molecules are too large to pass through the semi-permeable membrane) (1)

ii) the starch had been broken down/digested to sugar by the amylase (1)

d) as a control (1)

e) i) R (1)

ii) P (1)

f) he assumed that human amylase worked in the same way as the amylase solution he used (1)

3 Respiration, energy and exercise (page 22)

1 a) alveoli/air sacs (1)

b) carbon dioxide (1)

c) oxygen (1)

d) fingernails (1)

e) respiration (1)

f) 70 bpm (1)

g) emphysema (1)

h) asthma (1)

2 a) i) muscle (1)

ii) upwards and outwards (1)

b) heart; lungs (2)

3 a) i) type of food (1)

ii) the rise in the temperature of the water (1)

iii) any two of:

- same mass of food
- same volume of water
- same temperature of surroundings (2)

iv) a good answer would include at least two of:

- can calculate means
- results should be more reliable
- less emphasis on one 'bad' result (2)

    b) i)   respiration   (1)

       ii)   any two of:

- growth
- division of cells
- movement
- to maintain body temperature   (2)

4 a) green to red-orange   (1)

    b) some energy and lactic acid   (1)

5 a) i) $1000\,cm^3$ (1)

      ii)   Before exercise:

- 3 breaths every 10 seconds so 18 breaths per minute
- $500\,cm^3$ of air per breath so $9000\,cm^3$ of air per minute
- 20% of 9000 = $1800\,cm^3$ of oxygen per minute

During exercise:
- 5 breaths every 10 seconds so 30 breaths per minute
- $1500\,cm^3$ of air per breath so $45\,000\,cm^3$ of air per minute
- 20% of 45 000 = $9000\,cm^3$ of oxygen per minute

Difference:
$9000 - 1800 = 7200\,cm^3$ (4)

      iii) glucose + oxygen $\rightarrow$ carbon dioxide + water + energy   (2)

    b) the heart   (1)

    c) i)   less energy is released

          lactic acid is harmful   (2)

      ii) yeast   (1)

          alcohol (brewing) or carbon dioxide (baking)   (1)

6 a) i)   D   (1)

      ii) E   (1)

    b) one of:

- more mucus, so increased coughing
- damaged cells, so mucus tends to slip back into lungs, causing infection   (1)

    c) i)   oxygen   (1)

      ii) carbon dioxide   (1)

      iii) diffusion   (1)

      iv) thin lining means it is not far for gases to diffuse

          large surface area means many gas particles can cross at the same time   (2)

7 a) 1 g of fat contains 38 500 J, so 0.2 g contains $\dfrac{38\,500}{5} = 7700\,J$

      7700 J can raise the temperature of $25\,cm^3$ of water by $\dfrac{7700}{4.2 \times 25} = 73\,°C$ (3)

    b) any two of:

- some might be carried away on air currents (lost to the environment)
- some is lost as light energy
- some is used to warm up the glass and not the water   (2)

4 Reproduction in humans (page 26)

1 a) a sperm   (1)

   b) 28 days   (1)

   c) fertilisation   (1)

   d) testes   (1)

   e) embryo   (1)

   f) puberty   (1)

   g) a zygote   (1)

   h) placenta   (1)

   i) gestation   (1)

2 a) i)   E   (1)

       ii)   D   (1)

       iii) A or B or both   (1)

   b) i)   C (sperm duct/vas deferens)   (1)

       ii)   the cut tube prevents sperm from being ejaculated so they cannot reach the egg cell   (1)

   c) any two of:

       ● growth of facial hair
       ● deepening voice
       ● muscle development
       ● growth of pubic hair   (2)

3 a) i)   D   (1)

       ii)   B   (1)

       iii) C   (1)

   b) 9 months   (1)

   c) protection against damage/drying out   (1)

   d) an explanation should contain the following points in the correct order:

       ● digested in small intestine
       ● absorbed into blood
       ● transported in blood
       ● moves across placenta and along umbilical cord   (4)

4 a) i)   sperm   (1)

       ii)   tail for swimming

             nucleus carrying father's genes   (2)

   b) i)   fertilisation   (1)

       ii)   in the oviduct   (1)

   c) embryo; uterus; implantation; placenta/umbilical cord   (4)

5 a) ovaries; month/28 days; menstruation; blood; period   (5)

   b) i)   around days 13–15   (1)

       ii)   days 1–4   (1)

       iii) days 14–17   (1)

6 a) i)

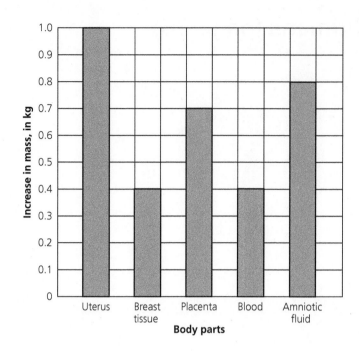

bars must not be touching and figure for fat must not be included (3)

    ii) 9 kg (1)

    iii) $\dfrac{0.7}{9.0}$ = 0.078 to 0.08 (2)

b) protein for growth of (fetal) tissues (e.g. muscle)

calcium for bones and teeth (for developing fetus as well as for woman) (2)

c) nicotine causes the baby to have a faster heart rate and the baby can be born addicted to nicotine

carbon monoxide causes less oxygen to be carried by red blood cells so the baby can have slower respiration and less energy for growth (2)

5 Reproduction in flowering plants (page 30)

1 a) produces pollen (1)

b) chloroplasts (1)

c) iodine solution (1)

d) cellulose (1)

e) stigma (1)

2 a) A – ovary; B – pollen grains; C – filament; D – anther; E – stamen; F – style; G – stigma (7)

b) any two from:

- small pollen grains – float on the wind easily
- anther at end of long filament – pollen held out into winds
- stigma is long and feathery – large surface to catch pollen (2)

3 a) A – NO [no water]; B – YES; C – NO [no oxygen in water]; D – YES (4)

b) i) check accuracy of plot (2)

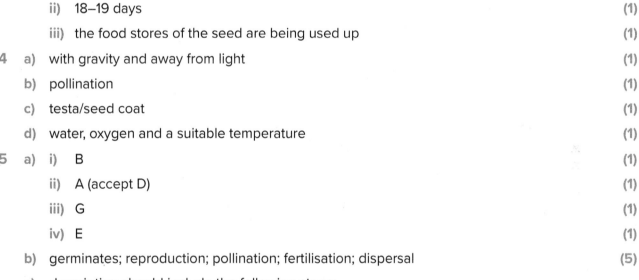

ii) 18–19 days (1)

iii) the food stores of the seed are being used up (1)

4 a) with gravity and away from light (1)

b) pollination (1)

c) testa/seed coat (1)

d) water, oxygen and a suitable temperature (1)

5 a) i) B (1)

ii) A (accept D) (1)

iii) G (1)

iv) E (1)

b) germinates; reproduction; pollination; fertilisation; dispersal (5)

c) description should include the following steps:

- grind up seeds – wear safety glasses (safety precaution)
- suspend ground-up seeds in water and then mix with iodine solution – don't get iodine solution on hands (safety precaution)
- look for blue-black colour
- water without ground-up seeds (control) to demonstrate that no unknown variable is affecting the results; this should not react with iodine solution and iodine solution should not turn blue-black, i.e. there is a colour change only if starch is definitely present (4)

6 a) increase the chance of pollination (1)

b) i) increased rate of oxygen uptake and increased temperature are related (proportional to one another) (1)

ii) oxygen is used in aerobic respiration, which releases energy; some energy is given out as heat, so temperature rises (2)

c) i) glucose (1)

ii) some description of photosynthesis – word equation for second mark (2)

6 Healthy living (page 34)

1 a) infectious (1)

   b) athlete's foot (1)

   c) addicted (1)

   d) COVID-19 (1)

   e) liver damage (1)

   f) an antibiotic (1)

   g) constipation (1)

2 a) more smoking increases risk of heart disease

      risk increases with age/length of time smoking (2)

   b) i) a good answer should include: less blood flow to heart muscle so less oxygen for respiration; heart muscle cells 'die' if they do not have enough energy from respiration (2)

      ii) high blood pressure means delicate blood vessels are more likely to be damaged, and a person is at higher risk of a stroke (1)

   c) i) needs a cigarette approximately every 100 minutes (1)

      ii) no cigarettes while asleep / nicotine levels drop below demand threshold while she is asleep (1)

      iii) at 14 mg per litre of blood (1)

      iv) will be reduced to approximately 50 minutes (1)

      v) nicotine patches / nicotine gum (1)

3 a) nerve impulses travel more slowly – reactions are slowed

      blood vessels close to the skin open up – person looks red-faced

      senses work less well – poor judgement of distance

      liver cells try to remove alcohol from blood – long-term liver damage (3)

   b) i) alcohol is absorbed through the stomach and small intestine into the pregnant woman's bloodstream and across the placenta to the fetus (3)

      ii) carbon monoxide causes less oxygen to be carried by red blood cells, so respiration slows and there is less energy for growth

         nicotine in smoke could make the baby become addicted (2)

4 a) DNA (1)

   b) typhoid (1)

5 a) any of: influenza/common cold/AIDS/COVID-19 (1)

   b) vaccination; antibody; antibiotic (3)

6 a) the possibility that some other infection could affect the patient (1)

   b) i) any two of: in food/drink; through the open wound from the air; on dirty bedclothes (2)

      ii) • wearing a surgical mask when in the patient's room – any infective organisms in the breath are trapped / cannot spread (1)
         • keeping the windows of the ward closed – prevents entry of infective organisms from outside the hospital (1)

   c) a diet low in animal fats; regular exercise; a lifestyle with little stress (3)

   d) making sure that children have new clothes (1)

7 a) i)   bacteria take time to multiply to harmful levels (1)

 ii) bacteria killed by antibiotic so levels fall (1)

 iii) bacteria hadn't been killed completely and multiplied again once Fumi stopped taking the antibiotics (1)

 b) i)   any one of:
- as part of blood plasma for transport
- helps in digestion
- part of fluid that lubricates joints (1)

 ii) acid removes enamel, exposing the vulnerable inner parts of the teeth to decay (1)

 c) vaccine stimulates body's defences (causes the body to produce antibodies); without actually giving the person the disease (2)

8 a) diet that contains all essential nutrients [award 1 mark if students name at least two of carbohydrates, fats, protein, vitamins, minerals]; in suitable amounts/proportions (2)

 b) i)   a disease caused by not having any/enough of a particular nutrient/vitamin/mineral (1)

 ii) lack of vitamin C [award 1 mark for lack of an essential nutrient found in fruit] (1)

 iii) sores that do not heal/bleeding gums/teeth falling out (1)

 iv) jam contains essential nutrient/vitamin C (1)

 v)   any (citrus) fruit (1)

7 Photosynthesis (page 39)

1 a) carbon dioxide (1)

 b) absorb minerals and water (1)

 c) chloroplasts (1)

 d) a cell wall (1)

 e) the sun (1)

 f) water and minerals (1)

 g) chloroplast (1)

 h) Chlorophyll (1)

2 a) i) $\text{carbon dioxide} + \text{water} \xrightarrow[\text{chlorophyll}]{\text{light}} \text{glucose} + \text{oxygen}$ (2)

 ii) iodine solution (1)

 iii) the straw-brown colour of the iodine would turn blue-black (1)

 b) B (1)

 carbon dioxide is necessary for photosynthesis; the atmosphere around leaf A has had all of the carbon dioxide removed by the sodium hydroxide solution (2)

 c) i)   some areas of the leaf are not green (do not contain chlorophyll) (1)

 ii) keep them in the dark for 24 hours (1)

 iii) leaf C – only the green area would be blue-black

 leaf D – only the area that is green and not covered by the black paper would be blue-black (2)

3 a) chlorophyll (1)

 b) pot has limited amount of compost, so minerals are soon used up (1)

c) i) potassium – C; phosphate – D; nitrate – A/C; magnesium – B (2)

ii) poisonous waste products are formed (alcohol or lactic acid)

less energy is released (2)

iii) either fungi or bacteria (1)

4 a) combustion (1)

b) break down waste materials (1)

c) respires only, using oxygen (1)

5 a) i) 0.18% (1)

ii) add carbon dioxide until concentration is 0.25%

more carbon dioxide means more photosynthesis but increase levels off after 0.25% so there is no point in increasing beyond this (2)

b) temperature; light intensity; type of plant used (3)

c) i) oxygen relights a glowing splint (1)

ii) collect gas over water in an inverted measuring cylinder (1)

iii) repeat the experiment and use mean values (1)

6 a) i) carbon dioxide (1)

ii) yellow (1)

iii) snails are respiring and releasing carbon dioxide (1)

b) i) purple (1)

ii) carbon dioxide has been removed as pondweed photosynthesises (plants also respire, but the photosynthesis exceeds the respiration, so the pH increases and the indicator turns purple) (1)

iii) balance between respiration and photosynthesis, so carbon dioxide levels did not change (1)

c) to show that the indicator did not change colour without respiration or photosynthesis (1)

8 Interdependence of organisms in an ecosystem (page 45)

1 a) break down waste materials (1)

b) photosynthesis (1)

c) a food chain (1)

d) iodine solution (1)

e) quadrat (1)

f) a carnivore (1)

g) green plant (1)

2 reproduction – leads to an increase in population

competition – reduces the number of organisms in a habitat

population – all the members of the same species living in one area

conservation – managing the environment for the benefit of wildlife (4)

3 a) i) • a herbivore – krill/squid

• a producer – phytoplankton

• a carnivore – any organism other than phytoplankton, krill or bacteria

• an organism that breaks down waste materials – bacteria (4)

ii) for example:

phytoplankton → krill → fish → crabeater seal → leopard seal (2)

b) any two of:
- streamlined shape
- webbed feet
- sharp beak (2)

4 a) the method used to count the organisms (1)

b) an animal (1)

c) woodland being cut down (1)

d) predators (1)

5 a) i) they were equal (1)

ii) line is horizontalb / the population size is steady (1)

b) i) C (1)

ii) rate of increase begins to slow (1)

iii) any one of:
- presence of predators
- disease
- availability of oxygen in water
- pollution (1)

c) i) protein (1)

ii) B (C is also acceptable) (1)

iii) the best profit can be made by selling the fish just before the population growth begins to level off; the population will continue to grow at the fastest rate if some fish are caught / removed from the lake (2)

6 a) hawks fall in number because fewer thrushes to eat

snails increase in number because there are fewer thrushes to eat them

lettuces fall in number because more snails eat them (3)

b) fewer thrushes so number of snails likely to increase

more snails so more lettuces will be eaten so fewer lettuces (2)

c) more nesting places so more babies

more possible foods so population will increase (2)

7 a) no trees to absorb water; water runs quickly off soil and downhill (1)

b) any two of:
- fewer nesting sites / less space for animals to live and breed
- fewer chances to feed / increased competition between animals for food
- fewer places to hide from predators (2)

c) i) fungi/bacteria (1)

ii) respiration (1)

iii) so that there are equal numbers of males and females for sexual reproduction (1)

iv) 33°C (1)

8 a) 5 ppm (no mark if no units) (1)

b) 2.6 km (allow 2.5 km) (1)

c) any one of: caddis fly larvae / freshwater shrimps / mayfly nymphs / stonefly nymphs (1)

d) rat-tailed maggots; *Tubifex* / sludge worms (2)

e) there are fewer insects/organisms for them to feed on (1)

f) long tube/tail can reach the air to obtain oxygen (1)

g) Answer should include ideas of

- sampling before conservation plan

- possible financial costs

- how to manage oxygen levels

- compromise between fishing and biodiversity.

Any sensible answers with sound biological background. (3)

9 Variation and classification (page 51)

1 a) have three main body parts (1)

b) nucleus (1)

c) has jointed limbs (1)

d) eye colour (1)

e) adaptation (1)

f) has a beak (1)

g) classification (1)

h) Blood group in humans (1)

i) photosynthesise (1)

2 angelfish (1)

3 a) i) Eric and Beth (1)

ii) the ones that lay the most eggs (1)

b) size of eggs; resistance to disease (2)

4 a) the type of skin it has

whether or not it has a bony skeleton

whether or not it has eyes (3)

b) spider – two body parts and eight jointed legs

insect – three body parts and six jointed legs

fungus – cells with a definite cell wall but no chlorophyll

fern – produces spores and cells contain chlorophyll

protist – body is made of a single cell, with a clear nucleus and cytoplasm (5)

5 a)

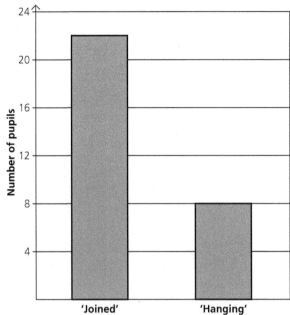

(2)

b) i) otherwise results could not be compared / measurements would not be accurate (1)

ii) cm (1)

iii) 3rd and 4th bars overlap, as do 4th and 5th bars (1)

(5)

6 a)

Name of mammal	Letter
Clethrionomys glareolus	C
Oryctolagus cuniculus	D
Sciurus carolinensis	E
Sorex araneus	A
Talpa europaea	B

b) presence of fur/hair

young drink/suckle milk (accept 'presence of mammary glands') (2)

7 a) otters can keep up their body temperature; so digestion speeds up and provides more soluble food for respiration which also speeds up to provide more energy; muscles are more flexible compared with the fish (3)

b) i) remain close to 40 °C, i.e. unchanged (1)

ii) rise to 25 °C (1)

c) i) any two of:

• streamlined shape

• webbed feet

• long, powerful tail (2)

ii) feed their young on milk (1)

iii) each has a backbone (they are vertebrates) (1)

d) variation (1)

Chemistry

10 Particle theory and states of matter (page 57)

1. a) vibrate more (1)
   b) evaporation (1)
   c) condensation (1)
   d) the independent variable (1)
   e) diffusion (1)
   f) $\dfrac{mass}{volume}$ (1)
   g) a liquid turns into a gas (1)
   h) remains the same (1)

2. a) boiling or evaporation (1)
   b) melting (1)
   c) freezing (1)
   d) condensation (1)

3. a) freezes below 0 °C (1)
   b) rise (1)

4. a) 34 cm3 – 30 cm3 = 4 cm3 (2)
   b) 124 g – 80 g = 44 g (1)
   c) density = $\dfrac{mass}{volume}$ = $\dfrac{44\ g}{4\ cm3}$ = 11 g/cm3 (2)
   d) the brooch is not pure silver, but it could contain silver as an alloy with lead (1)

5. a) correct positions of the three processes (3, 1 mark for each)
   b) liquid water – B; water vapour – A; ice crystals – C (3)

6. a) air-filled balloons are more dense than air, but helium-filled balloons are less dense than air (2)
   b) i) A (1)
      ii) C (1)
   c) i) diffusion (1)
      ii) particles of helium are smaller than those of the gases in air, so move more easily between the particles that make up the balloon wall (1)

7. a) i) pressure falls as individual molecules have less energy to 'press' against the surface area of the inside of the container (2)
      ii) likely to get smaller/weaker (1)
      iii) high temperature would increase pressure of the gas inside the can, so the can could explode (1)
   b) closer together than those in the gas; the same size as those in the gas; moving more slowly than those in the gas (3)

8. a) i) B (1)
      ii) temperature does not change as a substance changes state; temperature of the wax did not change between points B and C (2)
      iii) liquid (1)

b) any two of:

- readings can be taken more frequently (continuously)
- can go away and leave readings being taken
- likely to be more accurate (2)

c) water bath is used so maximum temperature is 100 °C (boiling point of water) (1)

d) (answer must refer to water and not to wax) water molecules moving more rapidly at end of experiment; leave surface as they gain enough energy from heat to evaporate away (2)

9 a) the ping-pong balls and balloon will start to move (1)

b) the ping-pong balls represent particles of water, and the balloon represents a pollen grain (2)

c) the water molecules moved around; they hit the pollen grains; the pollen grains moved around as they absorbed energy (3)

d) diffusion is caused by the random movement and collisions of particles; Brownian motion demonstrates how particles collide and move; in areas of high concentration, molecules collide more often; there will be fewer collisions if the molecules move apart, i.e. towards areas of lower concentration (2)

11 Atoms, elements and compounds (page 63)

1 a) is shiny in appearance (1)

b) an element (1)

c) moving slightly (1)

d) an oxide (1)

e) iron (1)

f) atoms (1)

g) molecules (1)

h) Fe (1)

i) 3 (1)

j) 21% (1)

2 a) aluminium in saucepans – it is a good conductor of heat

gold for a necklace – it stays shiny because it does not react with oxygen in air

mercury in a barometer – it stays liquid at room temperature

helium in a balloon – it is lighter than air

copper in cables – it conducts electricity and is easy to stretch (5)

b) i) any two of:

- good conductor of heat
- easily made into correct shape
- high melting point (2)

ii) it is a gas so can be compressed easily (into liquid form) (1)

3 a) it has a high melting point; it is a good conductor of heat and electricity (2)

b) filtration (1)

4 a) B: poor conductor of heat   (1)

   b) D: can be compressed   (1)

   c) F: very flexible   (1)

   d) E: good conductor of heat and G: very high melting point   (2)

5 a)

Weighing machine / Electronic balance   (3)

   b) i)   magnesium oxide   (1)

       ii) 4.4–4.5 g   (1)

       iii) 9.3–9.5 g   (1)

       iv) they are related, i.e. mass of oxide formed depends on mass of
            magnesium available   (2)

6 a) i)   A   (1)

       ii) E   (1)

   b) a compound contains only one type of molecule/a compound has a fixed number
       of atoms; a mixture has more than one type of molecule/a variable number of atoms   (1)

   c) i)   any of: oxygen (O_2) / hydrogen (H_2) / chlorine (Cl_2) / any other diatomic
            molecule (molecule with two atoms)   (1)

       ii) most likely answers are helium (He) and oxygen (O_2)   (2)

7 a) mercury   (1)

   b) oxygen   (1)

   c) Mg; Cl   (2)

   d) iron sulfide   (1)

   e) i)   calcium hydroxide   (1)

       ii) green-blue   (1)

   f)

Compound	Number of atoms of iron	Number of atoms of oxygen
FeO	1	1
Fe_2O_3	2	3

(2)

12 Mixtures, solutions, separation and solubility (page 69)

1 a) a compound (1)

b) a suspension (1)

c) evaporation (1)

d) decanting (1)

e) air (1)

f) contains particles of only one type (1)

g) a product of photosynthesis (1)

h) a compound (1)

i) a solvent from a solution (1)

j) condensation (1)

2 a) B (1)

b) A (1)

c) D (1)

d) C (1)

3 a) some substance in the stain is soluble in ethanol; so some of this substance dissolves in the ethanol and is no longer in the fabric of the trousers (2)

b) i) chromatography (1)

ii) red; dark blue; yellow; pale blue (1)

iii) two (two 'spots') (1)

iv) solute (1)

4 a) i) mass of salt (1)

ii) boiling point of solution (1)

iii) volume of water and type of salt dissolved in water must be controlled; starting temperature of water and room temperature have little or no effect on results (either of these is a correct answer) (3)

b) i) about 107.5 °C (1)

ii) the gradient of the curve is $\frac{9.0}{60}$ = 0.15 °C per g of salt

100 g salt will raise the boiling point by 100 × 0.15 = 15 °C

the boiling point of water is 100 °C; 100 + 15 = 115 °C (2)

5 a) i) change in colour of the water (1)

ii) raise the temperature (1)

b) evaporate the copper sulfate solution (1)

c) i) 14 g (1)

ii) X (1)

iii) X and Y (1)

6 a) i) kept same temperature; same volume of water (2)

ii) curry powder (1)

iii) $\frac{88}{32}$ = 2.8 (3)

b) the water is very hot, and solubility depends on temperature (1)

7 a) i) distillation (1)

    ii) evaporation then condensation (1)

  b) if liquid is water, anhydrous copper sulfate will turn from white to blue or cobalt chloride paper will turn from blue to pink; if liquid is pure water it will boil at 100 °C (2)

  c) i) greater than 20 °C (1)

    ii) heat has been transferred from the steam in the glass tube to the water (1)

    iii) it is cooled; particles get closer together/it condenses (2)

8 concentrated – a solution with many solute particles in a small volume of solvent

solution – a mixture of a solvent and a solute

solvent – the liquid part of a solution

solubility – the amount of a substance that will dissolve in a liquid

soluble – able to dissolve (3)

13 Chemical reactions (page 76)

1 a) the reaction being reversible (1)

  b) glucose (1)

  c) carbon dioxide + water (1)

  d) neutralisation (1)

  e) wood (1)

  f) adding extra water (1)

  g) Separation of a mixture of iron filings and sulfur (1)

2

Reaction	Use for the reaction
iron oxide + carbon monoxide → carbon dioxide + iron	to extract a metal from its ore
methane + oxygen → carbon dioxide + water	to transform chemical energy to thermal energy for heating
water + carbon dioxide → glucose + oxygen	for plants to make food
glucose + oxygen → carbon dioxide + water	to release energy in living organisms

(4)

3 a) i) iron + sulfur → iron sulfide (1)

    ii) loss of magnetic properties means that a new product has been formed (1)

  b) i) zinc sulfide; ZnS (2)

    ii) sulfur + oxygen → sulfur dioxide (1)

    iii) oxidation (1)

4 a) i) danger of damage to retina/eye (1)

    ii) to remove any magnesium oxide from the surface of the magnesium ribbon (1)

  b) i) 50; 62; 70.0 (1)

    ii) formation of magnesium oxide so a gain in mass; equivalent to the mass of oxygen consumed in the reaction (2)

    iii) magnesium + oxygen → magnesium oxide (1)

5 a) copper (1)

  b) sugar → alcohol + carbon dioxide + energy (1)

6 a) source of methane – paddy fields/waste gases from domestic animals/rubbish tips

    name of gas – carbon dioxide

    percentage overall contribution to the greenhouse effect – 56%   (2)

  b) any three of:

- stormy/extreme weather
- raised sea levels
- spreading of pests
- flooding
- formation of deserts   (3)

  c) i) the pH will be raised   (1)

    ii) calcium sulfate   (1)

    iii) calcium hydroxide + sulfuric acid → calcium sulfate + water   (1)

7 a) i)

Time, in minutes	Volume of resin mix, in cm^3
0	24
5	33
10	42
15	64
20	73
25	73
30	72

(2)

    ii)

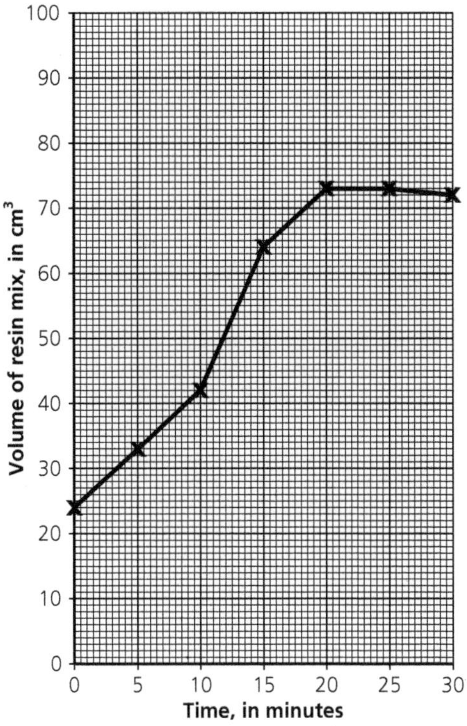

(3)

    iii) 14 minutes   (1)

  b) i) quantity of hardener   (1)

    ii) volume of resin mix   (1)

    iii) temperature; quantities of resin components   (2)

  c) they must not deteriorate when in the air, or when they get wet during use   (1)

8 a) the gas tap is fully on and the air hole fully open   (1)

   b) the blue part of the flame   (1)

9 a) because oxygen from the air is being used in the reaction   (1)

   b) boiling removes dissolved oxygen from water which would cause rusting without
       using oxygen from the air   (1)

   c) to the 17 cm³ mark   (1)

   d) air contains 20% oxygen, so the water level rose $\frac{2}{10}$ of the available air space   (1)

   e) heavier than at the start of the experiment   (1)

   f) because iron in the nail has combined with oxygen to form iron oxide, which is
       heavier than the iron alone   (1)

10 combustion; sulfur dioxide; sulfuric acid; acid rain; limestone   (5)

14 The reactions of metals (page 81)

1 a) silver   (1)

   b) decomposition   (1)

   c) increase   (1)

   d) Na   (1)

   e) usually of low density   (1)

   f) graphite   (1)

   g) iron   (1)

   h) Oxygen and water   (1)

2 a) i)   type of metal   (1)

       ii)   volume of gas released   (1)

       iii) by counting the bubbles   (1)

       iv)   same volume of acid; same-sized pieces of metal   (2)

   b) i)   no bubbles   (1)

       ii)   gold is very unreactive   (1)

3 a) i)   bar height = 0; because layer of oil prevents air reaching nail   (2)

       ii)   warmth had a bigger effect than salt – the nail in tube 3 (warm water only)
             rusted more than the nail in tube 2 (salt only)   (2)

   b) i)   acidic   (1)

       ii)   hydrogen   (1)

       iii) hold lighted splint over the top of the mouth of the test tube in which the gas is
             collected; hydrogen will produce a 'pop'   (1)

   c) i)   acts as a barrier and also as a sacrificial metal   (2)

       ii)   the layer of zinc would wear away   (1)

4 a) carbon monoxide + iron oxide → iron + carbon dioxide   (3)

   b) i)   4.0 – 0.8 = 3.2; $\frac{3.2}{4.0}$ = 80%   (2)

       ii)   does not corrode   (1)

	c)	i)	0.8%	(1)
		ii)	high-carbon steel	(1)
5	a)		a metallic element	(1)
	b)	i)	ores	(1)
		ii)	silver	(1)
	c)		iron + silver nitrate → iron nitrate + silver	(3)
	d)		gold is very unreactive so does not form an oxide (dull) with oxygen from the air; aluminium is more reactive	(1)

15 Acids, alkalis and neutralisation reactions (page 86)

1	a)		corrosive	(1)
	b)		neutralisation	(1)
	c)		neutral	(1)
	d)		a salt	(1)
	e)		a carbonate	(1)
	f)		carbon dioxide	(1)
	g)		salt + hydrogen	(1)
	h)		litmus paper blue	(1)
	i)		sodium hydroxide	(1)
2	a)		C	(1)
	b)		A; B; D	(1)
	c)		E; F	(1)
	d)		neutralisation	(1)
3	a)	i)	red-orange	(1)
		ii)	1–2	(1)
	b)	i)	hydrochloric acid + magnesium carbonate → carbon dioxide + magnesium chloride + water	(2)
		ii)	release of the gas carbon dioxide	(1)
		iii)	there was no more acid to react with the magnesium carbonate / all the acid had reacted	(1)
	c)		a salt; a compound	(2)

4 a)

	Acid or alkaline	Colour of indicator solution
Wasp sting	Alkaline	Blue
Bee sting	Acid	Red

(4)

	b)	i)	vinegar	(1)
		ii)	bicarbonate toothpaste or baking soda	(1)
	c)		helps to whiten teeth	
			helps to neutralise acidic foods, which are likely to damage teeth	(2)
	d)	i)	some leaves release an alkali, which neutralises the formic acid	(1)
		ii)	crushing releases alkali from the cells	(1)

5   a) i)   methane; air   (2)

         no change in colour of universal indicator   (1)

     ii)   carbon dioxide   (1)

         most alkali needed to return solution to neutral   (1)

     iii) neutralisation   (1)

   b) i)   copper + carbonic acid → copper carbonate + hydrogen   (2)

     ii)   bronze is an alloy containing copper; the copper reacts with the acid in the
         air; copper carbonate/copper salt is green   (2)

6   a) calcium carbonate + hydrochloric acid → calcium chloride + carbon dioxide + water   (3)

   b) loss of carbon dioxide   (1)

   c) no more fizzing   (1)

   d) green   (1)

   e) marble; limestone; chalk   (3)

   f) i)   hydrogen   (1)

     ii)   hold lighted splint over the mouth of the test tube in which the
         reaction is taking place; hydrogen will produce a 'pop'   (1)

7   a) it is very soluble in water; it is not poisonous   (2)

   b) carbon dioxide is released; these bubbles make the cake lighter   (1)

Physics

16 Energy resources and transfers (page 91)

1   A – 4; B – 6; C – 1; D – 2; E – 3; F – 5   (5)

2   a) joule   (1)

   b) hydro   (1)

   c) gravitational potential   (1)

   d) biomass   (1)

   e) elastic strain store of energy   (1)

   f) Tidal   (1)

   g) Biomass   (1)

3   a) i)   light; nuclear; thermal (accept nuclear and thermal in either order)   (1)

     ii)   electrical; kinetic   (1)

     iii) gravitational potential; kinetic   (1)

   b) advantage: light is free; disadvantage: system will not work if light intensity is
       low/at night   (2)

4   a) coal – generating electricity in power stations

       natural gas – heating and cooking in homes

       petrol – fuel for cars

       kerosene – aircraft fuel   (4)

   b) a non-renewable fuel will run out, because it cannot be replaced as quickly as it is used   (1)

	c)	i)	a fuel made originally by photosynthesis	(1)
		ii)	the sun / sunlight	(1)
		iii)	fossil fuels are reliable; fossil fuels are very concentrated stores of energy (either answer)	(1)
5	a)		elastic; kinetic; thermal; thermal; dissipated	(5)
	b)		more energy must be transferred from an elastic store of energy to kinetic energy to provide the extra light energy to make the torch brighter	(1)
6	a)		dissipation	(1)
	b)		light	(1)
	c)		energy cannot be created or destroyed but can be transferred from one store to another	(1)
7	a)	i)	sunlight	(1)
		ii)	energy from the nuclear store of energy in the Sun causes thermal gradients; air moves along these gradients, causing winds; these can then drive wind turbines	(2)
		iii)	renewable	(1)
	b)		coal; gas; petrol is also acceptable (derived from a fossil fuel)	(2)
	c)		wood; photosynthesis; chemical; machine; heat	(2)
8	a)	i)	walls	(1)
		ii)	loss through walls is 35% of 10 000, i.e. $0.35 \times 10\,000 = 3500$ J	
			so a 75% saving would reduce this by $0.75 \times 3500 = 2625$ J	
			$10\,000 - 2625 = 7375$ J	(3)
		iii)	conductor; air; sound	(3)
	b)		layer of flexible material prevents heat loss through the gap; cost of 'lost' heat would be greater than cost of draught excluder	(2)
9	a)		D, E, B, A, C	(2)
	b)		moving water just compresses the air rather than moving it through the turbines/slower waves do not produce enough air movement to overcome friction in the turbine	(1)

17 Energy and electricity (page 96)

1	a)	turbine	(1)
	b)	battery	(1)
	c)	graphite	(1)
	d)	TV monitor	(1)
	e)	generator	(1)
	f)	battery	(1)
	g)	thermal store of energy and emits light	(1)
	h)	joule	(1)
2	a)	advantages – any two of:	

- easy to transfer along power lines
- produces no waste when it is used
- easily transfers energy from one store of energy to another
- easy to control delivery

disadvantages – any two of:
- cannot be stored in large quantities
- transfer requires high voltages, so can be dangerous (2)

b) i) test each insulator in turn by attaching the clips to either end and measuring the current that flows using the ammeter; more efficient insulators allow less current to flow (2)

ii) any one of:
- battery always supplies the same power
- same clips and wires offer the same resistance
- same thickness of insulator samples
- same length of insulator samples (1)

iii) any sensible example, e.g. plastic for body of plugs to prevent electric shocks (2)

3 a) i) panels were not always facing towards the Sun to the same extent power output is higher if panels face directly towards the sun (2)

ii) 6 hours (8 to 14 on the graph) (1)

iii) straight line at maximum height from original graph (i.e. 52 kW), as shown (2)

b) they mean the satellites/space stations do not need to carry another form of fuel, which would be very heavy (1)

4 a) all correct for 5 marks, lose 1 mark for each incorrect label:

A – cooling tower; B – electricity output; C – generator; D – turbine; E – furnace (5)

b) i) in gigajoules per tonne: coal 39; oil 36; gas 51; nuclear 42 (4)

ii) any two of:
- smoke is harmful
- production of greenhouse gases
- coal is non-renewable (2)

iii) advantage – it does not produce pollutants

disadvantage – it is less reliable since winds are not consistent in strength (2)

5 a) chemical; kinetic; electricity; light; thermal (5)

b) the battery is able to store electrical energy, so the lights will stay bright even if the dynamo is not turning (e.g. when the engine is turned off) (2)

6 a) i) 58% (1)

ii) 36% (1)

iii) thermal store of energy (accept 'heat' as an alternative for 1 mark) (1)

b) i) lumo 17.3; glo-bright 16; eco-save 18; brite-lite 20; supa-glow 18.3 (1)

ii) bars must not be touching (3)

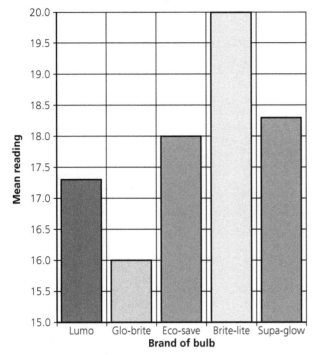

iii) brand of bulb (1)

iv) brightness of bulb (1)

v) same power supply; length of warm-up time; distance from sensor (3)

vi) repeated readings/checked each other's readings/calculated means (1)

vii) not all bulbs are equally efficient in conversion of electrical energy to light energy / not all bulbs offer the same value for money (1)

18 Forces and linear motion (page 101)

1 a) 14 km/h (1)

b) speed = $\dfrac{\text{distance}}{\text{time}}$ (1)

c) newton (1)

d) 72 N (1)

e) a size and a direction (1)

2 a) i) run 5 (1)

ii) $\dfrac{6.02 + 6.23 + 6.00 + 6.19 + 6.21}{5} = 6.13\,\text{s}$ (1)

iii) it reduces the effect of any one result, which could be unrepresentative (1)

b) i) speed = $\dfrac{\text{distance travelled}}{\text{time taken}}$ (1)

ii) $\dfrac{500\,\text{m}}{6.13\,\text{s}} = 81.6\,\text{m/s}$ (2)

c) the air resistance/wind resistance was different in the two directions (1)

3 a) X – upthrust; Y – gravity (2)

   b) friction of water

       friction of air/air resistance/drag (2)

   c) stabiliser angled downwards, i.e. front end below rear end (1)

4 a) i)   19 minutes (1)

       ii) between C and D (1)

       iii) B to C/E to F (1)

       iv) $\dfrac{4\,km}{19\,minutes} = 0.21\,km\,per\,minute$

            $0.21 \times 60 = 12.6$ km/h (2)

   b) i)   to increase friction between feet and mud/prevent slipping (1)

       ii) chemical; kinetic; thermal (1)

5 a) $7.5\,cm^3$ (1)

   b) all forces on it are balanced (1)

6 a) i)   it reduces wind resistance/air resistance/drag (1)

       ii) friction (1)

       iii) it has very little friction on the track surface (1)

   b) i)   200 m/s (1)

       ii) 700 m (1)

       iii) 600 m (1)

       iv) 12.5 s (1)

7 a) i)   4.4 N / the result at mass of 400 g (1)

       ii) 430 g; 2.4 N (2)

   b) i)   gravity (1)

       ii) mass; number; type (3)

   c) i)   gravitational force is lower in space so the forces needed to overcome
            gravity are also lower (award 1 mark for referencing lower gravity and award
            1 mark for referencing lower air resistance) (2)

       ii) arrow upwards diagonally to the right (1)

8 a) i)   liquids are denser than gases so take up much less volume for the same mass of fuel (2)

       ii) no oxygen in space / less/insufficient oxygen higher in atmosphere but plenty
            in Earth's atmosphere/near Earth's surface (1)

   b) i)   thrust (upwards); gravity (downwards) (2)

       ii) weight of rocket and fuel is greater than the thrust before 20 s (1)

       iii) upward force = 54 million N; total weight = 35 million N;

            resultant force = 54 – 35 = 19 million N upwards (2)

       iv) fuel is being burned and the products from burning the fuel are lost to the
            atmosphere as water vapour (2)

9 a) both springs extended to the same extent, so that the overall extension is twice
that of the original single spring (1)

b) both springs extended to the same extent, but the load is shared and each
one is extended by only half the extension of the original single spring (1)

19 Friction, motion and pressure (page 107)

1 a) newton/m² (1)

b) drag (1)

2 a) to increase friction between tyres and mud/prevent slipping (1)

b) i) to provide a small surface area so that pressure applied to the log will be very high (1)

ii) allow 1 mark for the correct number and 1 mark for the appropriate unit:

$$\frac{600N}{1.2cm^2} = 500 \text{ N/cm}^2$$

500 × 10000 = 5000000 N/m² (accept answer in N/cm² or N/m²) (2)

3 a) i) forcemeter/newtonmeter (1)

ii) to prevent the trolley sliding without any effort being applied (otherwise you may
not get any reading on the forcemeter) (1)

b) i) type of material (1)

ii) reading on forcemeter (1)

iii) any two of:

● added mass

● surface on which material is sliding

● mass of trolley

● forcemeter used to take readings (should always be the same) (2)

c) to allow the calculation of an average/mean (value to make results more reliable) (1)

d) i) P – 2.0; Q – 4.5; R – 3.2; S – 1.0; T – 6.5 (2)

ii) T (1)

iii) S (1)

iv) values would fall, as friction would be reduced (1)

4 a) i) increases surface area; which decreases pressure so there is less chance of falling
through crust of snow (2)

ii) bristles increase friction/prevent slipping

fur-like hairs provide thermal insulation/limit heat loss by conduction (2)

b) lubricant; friction (2)

20 Sound (page 109)

1 a) plucking the string harder (1)

   b) much more slowly (1)

   c) eardrum (1)

   d) listen to a message on the phone (1)

   e) cannot travel through a vacuum (1)

   f) amplitude (1)

   g) hertz (1)

2 a) i) the bell rings when the circuit is completed (1)

      ii) the sound gradually fades/disappears as the air is withdrawn from the bell

           sound cannot travel through a vacuum (1)

   b) i) sound travels more slowly than light (1)

      ii) less than one second (1)

3 a) i) B and C (1)

      ii) A and C (1)

   b) i) vibration causes the eardrum to vibrate; louder sounds make it vibrate so much
           it might tear or split (2)

      ii) 89 decibels is halfway between 88 decibels and 90 decibels, so time is
           halfway between 2 hours and 4 hours = 3.0 hours (Award marks for any answer
           between 2.6 and 3 hours. There is no requirement for mention of logarithmic
           relationship.) (2)

4 a) i) sound (vibrations) cannot travel through space/vacuum on Moon (1)

      ii) vibrations can travel through air in helmet and through solid material of helmets (2)

   b) i) solar energy (1)

      ii) by electricity to chemical store of energy (1)

   c) P (1)

5 a) the nerve to the brain is damaged (eardrum damage is temporary) (1)

   b) higher (1)

6 a) to prevent the bang tearing her eardrums/causing damage to her hearing (1)

   b) sound travels more slowly, so he would start his stopwatch later

       this would give an apparently shorter/quicker time for his runner (2)

   c) the spectators could have heard the sound as it travelled directly from the gun;
       as well as an echo from the dining-hall wall (2)

7 a) vibrations into electrical messages (1)

   b) i) more than 45 000 Hz but less than 65 000 Hz

           too high for robin and dog but not too high for cat (2)

      ii) the echo is reflected, so travels twice the distance

           thus the sound takes 0.1 s to reach the prey

           in 0.1 s sound travelling at 1500 m/s travels 150 m (3)

8 a) i) (label for material on *x*-axis (1 mark); sound level on *y*-axis (1); accuracy of bar heights (1); bars not touching (1))

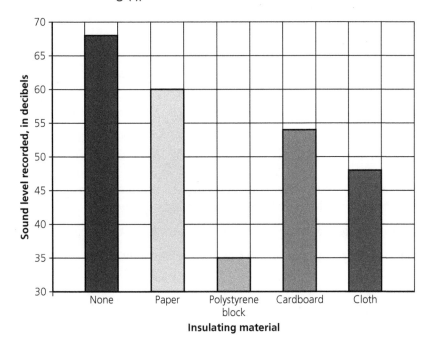

(4)

ii) polystyrene block   (1)

b) i) type of insulating material   (1)

ii) sound detected by sensor   (1)

iii) the box used and the distance between the sound sensor and the box   (2)

21 Light (page 115)

1 a) the Moon   (1)

b) opaque   (1)

c) much more quickly   (1)

d) in straight lines   (1)

e) in front of the mirror   (1)

f) incidence is equal to the angle of reflection   (1)

g) reflection   (1)

h) different frequencies   (1)

i) dispersion   (1)

j) sunlight being dispersed as it passes through water drops   (1)

2 a) so that light is only able to pass through the hole in the card and the star shape creates a clear shadow   (1)

b) lines pass through the gap in the card but only reach the screen by passing above or below the star shape (not through it)   (1)

lines must be straight drawn with a ruler and with arrows on them showing direction of ray (1)

c) the size of the shadow would decrease   (1)

d) translucent   (1)

3 **a)**

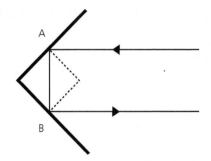

b) (parallel lines (1 mark); angle of incidence = angle of reflection, 1 mark for incident ray and 1 mark for reflected ray)

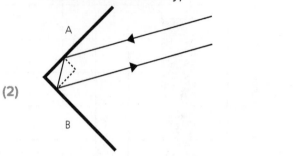

(2)

(3)

c) **i)** white light from headlight beam reflects from multiple surfaces; whatever the angle of incidence, some light reflects back to eyes of driver (2)

ii) light would only be reflected directly back to road users if it hit the mirror at 90° (1)

iii) the material has a roughened surface which does not reflect light in any organised way (1)

4 **a)** refraction (1)

b) some light has been absorbed (1)

c) red; orange; (yellow); green; blue; indigo; (violet) (4 if all correct)

5 **a)** mirrors at 45° within periscope tube (2)

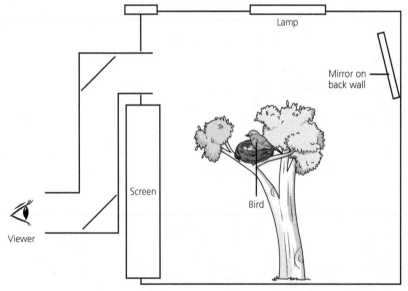

b) lines should be straight and with arrows showing direction of ray (1)

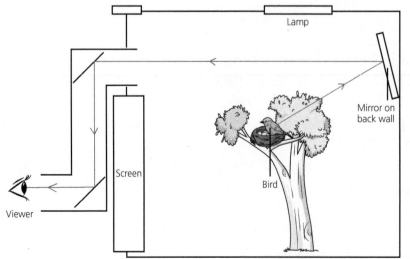

c) dispersion splits white light into different colours; this would make the image of the bird look multi-coloured around the edges **(2)**

6 a)

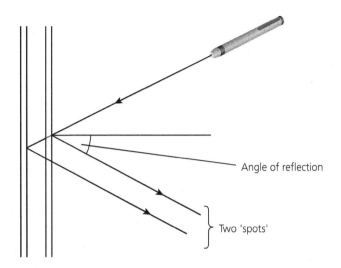

Angle of reflection

Two 'spots'

    i) ray to show reflection from front pane (award mark even if refraction within the panel is not shown) **(1)**

    ii) angle of reflection labelled **(1)**

    iii) rays to show reflection from back pane to produce second spot **(2)**

  b) some of the light is transmitted through the window **(1)**

       there are two reflections/the beam is split into two **(1)**

7 a) light from the overhead lamp is reflected from the smooth surface of the balls; and into the eyes of the player **(2)**

  b) light energy to electrical energy **(1)**

  c) light travels in straight lines and so the area behind each ball will not receive light **(1)**

  d) snooker balls are smooth so reflection will be 'clear' making snooker balls appear shiny; the table surface is rough, so reflection will be 'multi-angled' making table appear matt/rough **(2)**

22 Electrical circuits (page 120)

1 a) a chemical store of energy, is transferred by electricity **(1)**

  b) cell **(1)**

  c) switch **(1)**

  d) amp **(1)**

  e) a conductor **(1)**

  f) series circuit **(1)**

  g) a resistor **(1)**

  h) shine more brightly **(1)**

  i) series circuit **(1)**

2 a) A – switch; B – lamp; C – battery; D – ammeter; E – resistor   (5)

    b) i)   the battery   (1)

        ii) the ammeter   (1)

        iii)

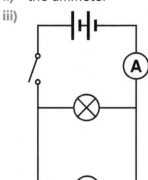

  (2)

    c) copper   (1)

3 a) i)   both must be closed / switch 1 and switch 2   (1)

        ii) no: (blower motor is in series with switch 1, and heater will work only if both switches are closed)   (1)

        iii) it will still work, although it will receive more current (so may become hotter) (award mark for saying that motor will remain 'on')   (1)

    b) 1 mark for each correct comparison (N.B. 'or words to that effect' applies to each)

Feature	Series circuit	Parallel circuit
Value of current in different places	The same	Not the same
Number of pathways that current can take	One	Several
Effect of one damaged component	All components stop working	Components in other parallel parts of the circuit will still work

4 a) i)   the circuit was completed when the switch contacts were pushed together   (1)

        ii)   (3)

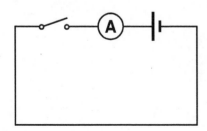

    b) B – bulb filament broken so circuit incomplete – replace bulb

        C – polarity of batteries is incorrect – reverse upper battery   (4)

5 a) (correct polarity of cell, 1 mark; correct position of ammeter, 1 mark)

  (2)

b) i)   length or thickness of wire   (1)

    ii) reading on ammeter   (1)

    iii) other wire features: e.g. if testing length should use same material and
         thickness; if testing thickness should use same material and length   (2)

c) take repeat readings for each wire and calculate mean/average   (1)

6 a) i)   lamps 1 and 2 will light up, but lamp 3 will not   (1)

       ii) 1.2A   (1)

       iii) 2.2A   (1)

   b) conductor; current; cell; lead; switch; switch; lamp; buzzer (1 for each 2 correct
       answers; no half marks)   (4)

7 a) E   (1)

   b)   (2)

Switch A	Switch B	Lamp X	Lamp Y	Lamp Z
Open	Open	Off	Off	Off
Closed	Open	**Off**	**Off**	**Off**
Open	Closed	**On**	**Off**	**On**

   c) all show 0.6A   (1)

23 Magnets and electromagnets (page 127)

1 a) poles   (1)

   b) iron   (1)

   c) coil   (1)

   d) magnetised   (1)

   e) C   (1)

   f) both poles   (1)

2 a) results: repels; nothing happens

       conclusions: X is steel; Y is the magnet; Z is tin (1 mark for identification of metal X;
       1 mark for correct result and metal Y; 1 mark for correct result and metal Z)   (3)

   b)

                                                                                                       (2)

   c) compass needle aligned with lines of force, pointing south   (1)

   d) magnetic field; attracted; north; repelled   (4)

3 a) current; material of coiled wire; number of turns in coiled wire   (3)

   b) i)   coil becomes an electromagnet; and so 'pulls' on the iron disc   (2)

       ii) 1.4 (glass has no effect as a core material)   (1)

4   a)   i)   same number of paper clips is picked up by each of them   (1)

    ii)   with iron as the core, the electromagnet can be switched on and off;

          steel stays magnetised even when the current is not flowing   (1)

    b)   when switched on, the electromagnet will attract iron and steel only, leaving the
         aluminium items behind; the iron and steel items can be dropped elsewhere when the
         electromagnet is switched off   (2)

    c)   i)   large current makes the electromagnet pull on the steel catch; this releases the
              contact, which pulls away because of the spring; circuit is now incomplete so
              current no longer flows   (2)

         ii)   it can be reset by pushing down on the reset button; an ordinary fuse
              must be replaced   (1)

5   a)   both force and direction   (1)

    b)   number of turns in the coil   (1)

    c)   A compass   (1)

    d)   A   (1)

6   a)   i)   the current may not have been great enough to cause the
              electromagnet to pull on the movable strip in the reed   (2)

         ii)   increase number of coils; increase current   (2)

7   a)   i)   increasing the current increased the strength   (1)

         ii)   any two of:

              • insert an iron core

              • increase number of turns in coiled wire

              • reduce resistance of coiled wire   (2)

    b)   circuit completed as switch closed; electromagnet 'pulls' on iron counterweight, lifting
         the barrier   (2)

8   a)   the switch is open so the circuit is incomplete; and the electromagnet cannot
         attract the armature   (2)

    b)   switch is closed → electromagnet pulls on armature → armature moves so gong
         strikes bell → contact breaker breaks circuit → spring pushes armature back again
         so circuit is completed once again   (5)

24 Space (page 134)

1   a)   150 kg   (1)

    b)   the Earth lies between the Sun and the Moon   (1)

    c)   star   (1)

    d)   lunar month   (1)

    e)   they reflect light from the Sun   (1)

    f)   mass × gravitational field strength   (1)

    g)   Venus   (1)

    h)   much faster   (1)

    i)   newtons per kilogram   (1)

    j)   light years (distance travelled by light in a year)   (1)

2 luminous; galaxy; telescope; light years; universe (5)

3 a) two straight lines drawn, with arrows to show direction; from transmitter to satellite and from satellite to receiver (2)

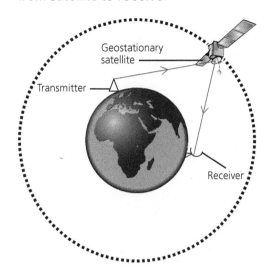

b) i) so that transmitted waves are always reflected to the receiver (1)

ii) the relative positions of the satellite and the dish will not change as long as the satellite is geostationary (1)

iii) one day/24 hours (1)

iv) gravity (1)

c) Moon; any planet (Earth most likely) (2)

4 1 year on Earth is 365 days – the Earth orbits the Sun

at the Equator, there are 12 hours of light and 12 hours of darkness – the Earth rotates on its axis

in Britain there are four seasons in the year – the Earth's axis is tilted

there is a new Moon every month – the Moon orbits the Earth

a ship sailing away from land goes out of sight – the Earth is a sphere (5)

5 a) i) D (1)

ii) B (1)

iii) A (1)

b) i) all arrows pointing towards centre of the Moon (1)

ii) collecting bag on chain always pointing towards the centre of the Moon (1)

c) i) the Earth orbits the Sun because the Earth is moving too quickly for gravity to pull it in to the Sun and too slowly to fly off into space (2)

ii) 1 year/365 days (1)

iii) $\dfrac{\text{distance}}{\text{speed}} = \dfrac{149\,000\,000\,\text{km}}{300\,000\,\text{km}/\text{s}} = 496.7\text{s}$

$\dfrac{496.7\,\text{s}}{60\,\text{s}} =$ just over 8 minutes (8.3 minutes) (2)

6 a) i)   it is in an orbit in a different plane from the other planets   (1)

    ii)   planets often have natural moons; Charon is acting like a moon around the
          planet Pluto   (1)

7 a)   any one incorrect label loses 1 mark   (2)

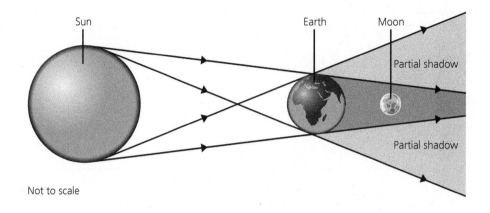

Not to scale

    b)   (no marks lost if partial shadow is not labelled)   (3)

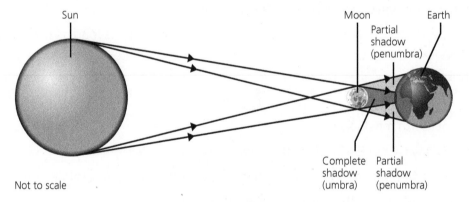

Not to scale

    c)   the darkness during a solar eclipse makes the birds think it is night, when they would
         usually stop singing   (1)